SURPRISE
THE UNION OF QUANTUM PHYSICS, RELATIVITY, AND
THE BIBLE

SURPRISE
THE UNION OF QUANTUM PHYSICS, RELATIVITY, AND
THE BIBLE

MARK HICKS

© 2012 by Mark Hicks. All rights reserved.

WinePress Publishing, PO Box 428, Enumclaw, WA 98022, functions only as book publisher. As such, the ultimate design, content, editorial accuracy, and views expressed or implied in this work are those of the author.

No part of this publication may be reproduced, stored in a retrieval system, or transmitted in any way by any means—electronic, mechanical, photocopy, recording, or otherwise—without the prior permission of the copyright holder, except as provided by USA copyright law.

All Scripture quotations, unless otherwise indicated, are taken from the *New American Standard Bible*, © 1960, 1962, 1963, 1968, 1971, 1972, 1973, 1975, 1977, 1995 by The Lockman Foundation. Used by permission.

Scripture quotations marked AMP are taken from the *Amplified Bible*, Copyright © 1954, 1958, 1962, 1964, 1965, 1987 by The Lockman Foundation. All rights reserved. Used by permission. (www.Lockman.org)

Scripture quotations marked CEV are taken from the *Contemporary English Version*. Copyright © 1995 by American Bible Society. Used by permission.

Scripture quotations marked ESV are from *The Holy Bible, English Standard Version*® (ESV®), copyright © 2001 by Crossway Bibles, a publishing ministry of Good News Publishers. Used by permission. All rights reserved.

Scripture quotations marked KJV are taken from the *Holy Bible, King James Version*.

Scripture quotations marked MSG are taken from THE MESSAGE. Copyright © by Eugene H. Peterson, 1993, 1994, 1995. Used by permission of NavPress Publishing Group.

Scripture quotations marked NKJV are taken from the *New King James Version*. Copyright © 1982 by Thomas Nelson, Inc. Used by permission. All rights reserved.

Scripture quotations marked NLT are taken from the *Holy Bible, New Living Translation*, copyright © 1996, 2004, 2007. Used by permission of Tyndale House Publishers, Inc., Carol Stream, Illinois 60188. All rights reserved.

ISBN 13: 978-1-60615-081-8
ISBN 10: 1-60615-081-2
Library of Congress Catalog Card Number: 2010936883

Dedicated to my wife, Bonnie, whose incredible patience, faith, love, and invaluable assistance made these words possible.

Special thanks to Paul Chapinduka, Steve Swanson, John Perry, Rick Johnson, Rev. Otis G. "Dad" Clark, Shirley O'Dell, Shirley Lamb, Dana Moran, Jeff Thomas, Ron and Bonnie Nelson, and Nathaniel Nelson for the inspiration.

"The most miraculous thing is happening," my visitor proclaimed with a painful sincerity, probably over rehearsed. "The physicists are getting down to the nitty-gritty, they've really just about pared things down to the ultimate details, and the last thing they ever expected to happen is happening. God is showing through. They hate it, but they can't do anything about it. Facts are facts. And I don't think people in the religion business, so to speak, are really aware of this—aware, that is, that their case, far-out as it's always seemed, at last is being proven."
Conversation between a computer student and a professor at a divinity school, in *Roger's Version* by John Updike[1]

I tell you the truth, you can say to this mountain, "May you be lifted up and thrown into the sea," and it will happen. But you must really believe it will happen and have no doubt in your heart.
—Jesus of Nazareth, Mark 11:23 (NLT)

www.realight.org

CONTENTS

Introduction...xi

1. Modern Science and the Bible Say the Same Thing?!.......... 1
2. Some Forgotten Corner of the Universe?.................. 3
3. Perspective.. 5
4. The Problem of Quantum Physics...................... 7
5. The Bible and "Common Sense"....................... 9
6. Is Imagination More Important Than Knowledge?.......... 11
7. The Mainframe.................................... 15
8. *Enterprise*....................................... 17
9. Nothing Holding It Up?............................. 20
10. Constant Motion.................................. 26
11. Our Own Point of View............................. 28
12. Gravity.. 30
13. "Universal Constants"?............................. 34
14. What's Really Out There?........................... 36
15. The Uncertainty Principle........................... 41
16. Light Seed....................................... 46
17. A Matter of Time.................................. 50
18. The Ultimate Question............................. 52
19. The Three Fundamental Forces....................... 55

20. Light Carriers...59
21. Falling Apart..63
22. Who Knew?..67
23. Star Stuff..72
24. Precisely Balanced..................................74
25. The Quest to Understand Light....................76
26. The Essence of Life.................................78
27. Nothing Stays the Same............................82
28. The Harmony of the World........................86
29. Quantum Leap......................................90
30. The God Particle...................................92
31. The Experiment That Won't Go Away.............94
32. The Light Is All....................................97
33. Searching for Aether...............................99
34. The Only Universal Constant....................102
35. The Fundamental Things Apply..................105
36. The Effects of Relativity..........................112
37. The Interchangeable Nature of Energy and Matter.........114
38. Does Anybody Really Know What Time It Is?...........118
39. The Paradox of Light.............................123
40. Standard Physics..................................128
41. No Explanation...................................133
42. Cause and Effect..................................135
43. An Observer-Created Reality....................139
44. Virtual Particles and Probability Waves..........146
45. Imagine..150
46. Space and Time Are Not Impassable Barriers....152
47. Good Vibrations..................................155
48. Two Basic Premises...............................160
49. Our Creative Power...............................162
50. Seek and You Will Find..........................164
51. The Information of Love........................167
52. The Greatest Truth...............................172

Endnotes..175

INTRODUCTION

> Still I look to find a reason to believe.
> —Tim Hardin

THIS IS A book about perspective: the way both science and the Bible view the world and universe we call home. Ultimately, however, it is about our own perspective: the way each of us sees what is all around us every day. But before we start to explore, it might be helpful to first take a look at the means by which we do that. How do we process what we see? What does our brain do with the information we take in? How do we try to make sense out of this crazy place?

When a fly looks at a TV screen it sees individual pixels. When a baby first opens his eyes, his initial perception of TV is not much different. That, however, lasts about as long as it takes him to blink and look again. From the very first day we perceive our surroundings, we begin to look for recurring patterns and attempt to categorize and assign meaning to them. All the fly will ever see is pixels. Those same pixels will soon represent Barney, Muppets, and an increasing myriad of stuff the child will almost immediately perceive Mom and Dad should buy.

Science defines consciousness as the ability to create new or higher levels of meaning out of a series of patterns or symbols. The way we do this shapes what we become. From the moment we are born, we spend our lives trying to understand and successfully relate to everything we sense around us. One of the first conclusions we reach is that the better we are able to do this, the happier we will be. When we figure out how to get our mother's attention, we quickly realize there are a lot of benefits associated with being able to communicate with her. There is nothing like the look on a child's face when he realizes that when he says "coook-eee," he receives one. You can almost see the light go on when it occurs to him that if he could get the hang of this, the sky could be the limit. This drive to understand our environment never changes, and the more accurate our comprehension of the people and things in this world, the more successful and the happier our time on it will be. Understanding brings harmony and peace to our lives. Misunderstanding brings disharmony and conflict.

The way we make sense of our environment shapes what we become. From the very first day we open our eyes, we begin to absorb all that is around us, look for recurring patterns, and attempt to categorize them. Children often go through a stage in which they announce just about everything that is going on. What they are doing is establishing and confirming their perception of reality.

As we learn new relationships, we store them for future reference. Another way to say this is that we are creatures of habit. This is a good thing, because it means that after continually being exposed to the same pattern, our brains relegate it to our subconscious so that our minds can be free to do other things. For instance, when you first started driving, you had to focus all of your faculties on just getting from one end of the driveway to the other without hitting anything with Dad's car. After gaining some experience behind the wheel, however, you no longer had to think about the location of the brake or accelerator or how much pressure to apply to them. Because of the continuous repetition, the patterns became so familiar that you were eventually capable of driving somewhere with your mind so involved in other things that you were hard-pressed to remember the trip. When you first sat at the computer keyboard, it was hunt and peck for just one word. Now you don't need to consciously think about it.

We are able to function at a high level because once we learn a pattern (e.g., driving, typing, playing a musical instrument), that knowledge is relegated to the subconscious, leaving our minds free to concentrate on higher functions such as communication and creativity. There is, however, a downside to this. If we come across a new pattern that has a lot in common with one we have repeatedly seen in the past, sometimes we may mistakenly assume the new pattern is the old one we already know so well. For instance, take a look at the patterns below:

Because we live in a three-dimensional world, it is hard for us to realize that the lines above are nothing more than a two-dimensional image and that the center lines are actually the same length. Go ahead and measure them. When we go to the movies or watch TV, the same thing happens. It never enters our mind that what we are seeing has no actual depth. Because of the patterns we have grown accustomed to seeing, we subconsciously fill in information that is not actually there and assume that the action is taking place in the 3-D world we know.

Our brains constantly scan our environment, attempting to construct a picture of reality that conforms to the patterns it knows—that we have on file. How many times have you seen someone in a grocery store whom you assumed was one person but, on closer inspection, turned out to be someone else? When you saw that person and believed he was someone you knew—only to find someone you did *not* know staring back at you—the same thing happened as when you looked at the figure above. Your mind scanned the stored patterns in your brain and came up with a match. It was only when you got a little closer that you discovered it was not the nose, cheekbone, or eyes that you originally thought you saw. Because you—quite necessarily—are a creature of habit, your brain had actually filled in some missing information that you had not really seen.

Given that our brain stores patterns in this manner—so that we do not have to continually use all of it to do everyday things—sometimes we make mistakes. This is true not only of everyday occurrences; it also affects the beliefs we form about ourselves, our lives, our reality—everything. When forming opinions about our environment, we take in the available information and scan or compare it to patterns we have on file. We absorbed and stored those patterns because past experience led us to conclude they were accurate representations of our reality.

Now misidentifying a pattern, such as a person we thought we knew in a grocery store, happens to everyone from time to time. That is not a major issue and can be quickly corrected. The problem occurs when these types of mistakes are not corrected. The more we misidentify patterns about the nature of our environment and store that information in our brains, the harder it will be for us to fix the problem so that we can successfully relate to our surroundings.

For example, if a person has negative self-image patterns imbedded within her at a young age, as an adult she will match current life experiences with those foundational negative patterns. Because that individual does not believe she has any personal worth, she will have a hard time accepting that she can do anything of value. She will refuse a job promotion because she won't believe that she can do it. She may interpret unfamiliar actions of others—even family members—as being hostile or critical. Of course, as with the figure above and the example of the misidentified person in the grocery store, the truth can often be quite the opposite. However, because she has believed an incorrect pattern to be true for so long, it will be extremely difficult for her to accept the possibility that she might be wrong.

The more ingrained a pattern is, the harder it is to alter. The longer we have held a particular view, the harder it is for us to admit we have been wrong all that time. Pride is a tough taskmaster.

An even greater problem occurs when someone misidentifies a pattern and passes that misinformation on as truth to others. In some cases, patterns misidentified by one person have been passed on for thousands of years and accepted as truth by millions of people. In such cases, it can be extremely difficult for individuals and even societies to recognize the error and make corrections that will benefit themselves and their world.

To illustrate this, imagine that you are living four hundred years ago. You and every person you know have always believed that the sun, the moon, and the planets circle around the earth. To you, this is such an obvious truth that you have, quite literally, never given it a second thought. Why would you? Just open your eyes and look around. So when someone like Galileo comes along and tells you that the earth is moving and the sun is stationary, all you have to do is walk outside, stand still, and not feel any movement to say, "Sure, pal, it will be okay. There are nice homes where people like you can go and not hurt themselves." History plainly shows us that just because we as individuals and as a society have concluded that certain patterns are correct for long periods of time, that does not make them so.

The patterns that each of us use to identify what we believe about the basic nature of the reality we inhabit are sometimes referred to as "core beliefs." These core beliefs are the ones we use to filter new information that relates directly to our understanding of who we are and how we interact with our environment. They govern the quality and quantity of our life and shape our choices about such things as our career paths, our friends, whom we might marry, and other important life decisions.

Because these core beliefs form the foundation on which we base our personal identity as well as our understanding of the world around us, we do not choose them lightly. For this reason, before we will discard one and accept something new, we need a good reason to do so. In the example above, before people could accept Galileo's theory, they needed proof. Galileo soon discovered that it wasn't easy for people to abandon their old beliefs and accept what he had seen through his telescope.

However, because our ability to successfully relate to our surroundings and navigate our way through this life is dependent on the accuracy of our core beliefs, if the reason is sound, we will accept the new view to be true regardless of the strength of our desire for it not to be so. For instance, at a funeral, we will not believe a loved one is still alive regardless of how strong our desire is for the opposite to be true. This may seem obvious—and a bit crude—but it brings home the point that *desire alone cannot alter a belief.* All of us believe many things that we wish we didn't.

Of course, if we are uncomfortable with a new view, we will often search long and hard for reasons to reject it. However, because our core beliefs govern our ability to successfully live on this planet, we will use the method of perceiving and processing external information through our previous experiences to alter them. So, although internal emotional desires certainly can, and do, lead to erroneous beliefs, these desires alone cannot create them. We must go through the process of finding a distinct reason to believe something before we will actually accept it.

A common perception among those who do not believe in God is that those who do came to their belief by means of some other process. For example, if a person suddenly becomes a Christian, a non-believer may assume that the person—maybe subconsciously—is responding to some form of trauma and has accepted a falsehood in order to gain some measure of peace in his or her life. My own opinion was that those who believed in God "chose" their belief as a weak-minded and superstitious response to the aspects of existence over which they had little or no control. Because I did not personally see any evidence for their belief, I concluded that such evidence did not exist and that they had simply convinced themselves of their ideas through concentrated wishful thinking. I thought that their beliefs were the result of ignorance and a corresponding overwhelming fear of the unknown. To put it plainly, I thought they were psychotic.

It was not long, however, before I found myself involved in—supposedly—intellectual discussions concerning the possibility of supernatural phenomena in "non-religious" scenarios. At this point, a friend asked if I thought it would be wise to exclude traditional metaphysical ideas from our inquiry. She suggested that a terse dismissal of some of these ideas due to our own personal experiences with "church" might be shortsighted on our part and, at best, was an inadequate reason to reject them. Although these ideas did appear childish to us, she was of the opinion that it would be unfair not to consider them along with all the other concepts swirling about. This led to another apparently fair thing to do: instead of flatly assuming that all "born-again" Christians were crazy, we should at least talk to one or two and ask them how they arrived at their beliefs.

What I found was that those who said they were born-again arrived at their belief the same way that people arrive at any other belief: they chose to consider the possibility that certain spiritual principles could be true, acted on that hypothesis, and let the results speak for themselves. Based on the information they received, they came to the conclusion that they were able to sense that God is real. The Bible says our capacity to sense what lies beyond our five natural senses is due to our "spirit" (see 1 Cor. 2:11, Prov. 20:27, James 2:26). Other common terms for this ability are extrasensory perception (ESP), paranormal activity, or telepathy. Most of us can sense the presence of another person without actually seeing or hearing him or her approach from behind.

Beliefs are never the result of desire alone. If they were, all of us would be living decidedly different lives. Rather, beliefs result from the processing of external information. However, the manner in which we process that ever-increasing amount of information has left many of us in a quandary when it comes to the scientific and the spiritual. There are two good reasons for this: (1) both science and religion relate to our core beliefs—the fundamental nature of what and who we are, and (2) the information that we have been given from both scientific and religious sources in recent times has been anything but consistent. In fact, it has been highly contradictory and often quite antagonistic.

It was at the turn of the twentieth century that Albert Einstein first came up with his special and general theories of relativity. Soon after, physicists such as Niels Bohr,[2] Werner Heisenberg, and others uncovered the nature of the quantum world in which atoms dwell. Most, if not all, physicists and scientists would agree that the impact of these theories on our understanding of our universe could be compared to few events in human history. Like Magellan's circumnavigation of the globe or Isaac Newton's explanation of gravity, the knowledge of relativity and quantum physics changed everything we thought we knew about our surroundings and unlocked undreamed-of possibilities.

The theories of relativity and quantum physics have done so much to improve the condition of humankind during the past one hundred years that there is no question they contain a large amount of truth. The proof "is in the pudding." Today, due to our knowledge of relativity and the quantum world, we have television, computers, cell phones, and an entire technological world that we simply take for granted.

Unfortunately, this form of scientific truth has often been assumed by those in the religious community to be at odds with the truth found in God's Word. In fact, many theologians and religious scholars, steeped in half a millennium of scientific and religious dichotomy, paid little attention to the discoveries of men like Einstein and Bohr at the time. That was a mistake, just as it was a mistake for the scientific community to assume that those who held religious beliefs had abandoned logic.

Many advocates on both sides of this issue seem to take more time criticizing the other than they do addressing the actual facts. As a result, we find ourselves uncomfortably lost somewhere in the middle. While we cannot deny what science has shown to be real—the facts speak for themselves—we also cannot deny that it is unrealistic to assume that there is nothing more to life than what meets our five physical senses. Yet neither side in this debate seems to leave any room for the other. The result is a lifetime of exposure to these confrontational patterns that has left an uncomfortable and conflicting uncertainty inside nearly all of us.[3]

This apparent contradiction between the spiritual and the physical did not always exist. In fact, before 1616, when Galileo Galilei began to teach his ideas about a sun-centered universe, science and religion had always been seen as complementary; each provided proof and insight into the other. That ended when Galileo opened his mouth. Church leaders condemned his ideas as "false and contrary to Scripture" and warned him to abandon them—which he promised to do. When he later defended his theory in 1632, the church took action to silence him. That little spat produced seeds of distrust and caused a split between religion and science that, over the centuries, have grown into an ever-increasing chasm. Today, the scientific and religious communities view each other as a threat to their pursuit of the truth.

Yet regardless of whether one is a theologian, evolutionist, atheist, born-again Christian, or someone who doesn't much care one way or the other, the truth will always remain what it is despite who it is that seeks it. This being the case, it would stand to reason that whatever amount of truth there is residing in the current state of scientific theory, if the Bible is true, it should be in agreement. And wouldn't you know, *it actually is*. In fact, the theories of modern physics confirm what the Bible has always maintained about the nature of God and His creation. The fundamental principle underlying both is the same.

Perhaps Albert Einstein—one of the few individuals in the past century who managed to be a cool voice amidst all the hot air—summed it up best:

> We are in the position of a little child entering a huge library filled with books in many languages.
>
> The child knows someone must have written those books. It does not know how. It does not understand the languages in which they are written. The child dimly suspects a mysterious order in the arrangement of the books but doesn't know what it is. That, it seems to me, is the attitude of even the most intelligent human being toward God. We see the universe marvelously arranged and obeying certain laws but only dimly understand these laws.

In Luke 10:21, Jesus, "filled with the joy of the Holy Spirit," said, "Oh, Father, Lord of heaven and earth, thank you for hiding these things from those who think themselves wise and clever, and for revealing them to the childlike. Yes, Father, it pleased you to do it this way" (NLT).

So with awe, wonder, humility, and a childlike appreciation of the world and universe in which we find ourselves, let's explore.

CHAPTER 1

MODERN SCIENCE AND THE BIBLE SAY THE SAME THING?!

"*What!?* That's ridiculous. Who are you trying to kid?"

NO, I AM not reading your mind; I am quoting my own initial reaction to this idea. Sometime later, however, I discovered a flaw in my reasoning: I really had no idea whatsoever about what modern science has to say about the basic nature of the reality we inhabit. In fact, when I asked around, I was hard pressed to find anyone who did. The reason, I discovered, was that scientists themselves are having a very hard time explaining the nature of the universe. Many incredibly brilliant people appear to be at a loss when asked for a rational explanation of how and why the quantum world of atoms behaves the way it does. I don't know about you, but I am also at a loss when asked such questions. So if your first reaction to the subject of this book is to say, "I don't think I understand these things," let me congratulate you. You are in the same boat with not only me but a great many physicists as well.

In fact, as you read through these pages, you may find that you know a lot more science than you thought you did. Think about it: when you are behind the wheel of your car, you are constantly calculating some very complicated geometric problems. You are also performing other calculations involving force, acceleration, and momentum. Amazingly, the values in these equations keep changing by the second, yet you are able to process them all as you take in the scenery and immerse yourself in the music flowing from the speakers. Just because you don't know how to transfer these concepts onto paper using a bunch of scientific hieroglyphics does not mean that you are unfamiliar with them. The fact that you are in a reasonable semblance of one piece at this moment is conclusive proof that you are an unmitigated master of these subjects. If you weren't, your obituary would have been written a long time ago.

So, being the intelligent person you have already conclusively proven yourself to be, it follows that just because you may not have understood what some teachers, professors, or people on TV have said about physics in the past, doesn't necessarily mean that the problem was with you. Having personally endured a good deal of higher education, one thing I found out is that people who consider themselves "intellectuals" tend to use words or symbols that only others in the profession can understand. After spending a lot of time and money to learn these terms, I believe I have finally come to understand the reason they do this. If the Board of Regents realized they were just saying something that every single person already knew, they would be out of a job or performing another possibly more valuable task for the institution.[4]

This language barrier between intellectuals and the rest of the population is, however, finally breaking down. The past few decades have seen a breakthrough in the ability of everyday folks to understand what intellectuals call "theology." Bible translators have taken what previously required years of study to properly understand (the King James Bible) and produced a refreshing number of contemporary language versions. Likewise, the past few years have seen a growing number of scientific publications attempting to deal with what had been the virtually obscured mysteries of quantum mechanics.

As you are well aware, the Bible speaks of realities that are beyond the abilities of our five senses to detect. Remarkably, the past decade has produced a continually growing amount of scientific literature probing the exact same thing.

CHAPTER 2

SOME FORGOTTEN CORNER OF THE UNIVERSE?

> It suddenly struck me that that tiny pea, pretty and blue, was the earth ... I didn't feel like a giant. I felt very small.
> —Neil Armstrong

THE UNIVERSE, AS we know it, is a giant black void. In this huge empty space are tiny pinpricks of light sparsely scattered like a few leftover grains of salt on a large dining room table. They are so far apart that from our little pinprick we call the sun, it would take more than fifty thousand years to journey to our closest neighbor, the star called Proxima Centauri, traveling at the fastest speed of any spacecraft we have ever launched[5]—not to mention the same amount of time that it would take to get back. Scientists are working on newer modes of space travel—such as ion propulsion—but even if these pan out as hoped, it will still take thousands of years to travel to the nearest star.

No one has seen the end of the universe. Nevertheless, scientists have seen quite a lot, especially since the Hubble Space Telescope went into orbit in 1989. One thing they have noticed is that although the closest star is more than twenty-five trillion miles away, there are about two hundred billion other stars grouped together with us in our little corner of the universe. We call this group the Milky Way Galaxy. The Milky Way, however, is just one of billions of other galaxies, each with billions of stars. Even so, our galaxy is quite large. In fact, if our solar system—the sun and the eight planets in orbit around it—were the size of a penny, the Milky Way would be larger than the states of California and South Carolina put together.

If you can, pick up just one grain of sand and take a look at it. One hundred billion of them would fill a dump truck. We define our universe as everything in existence: all of space, all time, and all of the matter and energy in it. In relation to the entire known universe, our galaxy—and its two hundred billion suns—is not a mere drop in the bucket; it is but a single grain of sand in the dump truck. There are at least two hundred billion other galaxies. As far as the number of stars in the universe is concerned, trying to count them all would *not* be like trying to count grains of sand because you would run out … of sand. There are more stars in the universe than there are grains of sand on all the beaches in the world. Carl Sagan[6] put it this way: "We find that we live on an insignificant planet of a humdrum star lost in a galaxy tucked away in some forgotten corner of a universe in which there are far more galaxies than people" ("Carl Sagan." BrainyQuote.com. Xplore Inc, 2012. 10 May. 2012. http://www.brainyquote.com/quotes/authors/c/carl_sagan_2.html).

CHAPTER 3

PERSPECTIVE

Few are those who see with their own eyes
and feel with their own hearts.
—Albert Einstein

HAVE YOU EVER been sitting in a car at an intersection and thought you were moving, only to realize that it wasn't you but the car next to you? In a situation like this, what you did to confirm that you *were not* in motion was take a quick look around, gather some additional information, and then alter your assessment of the situation based on your new perspective. This kind of readjustment has been going on since the dawn of man.

It is said that the ancient Chinese built towers to reach the moon and stars. Likewise, the Bible tells of a construction project in a city called Babel that included a tower "whose top may reach unto heaven" (Gen. 11:4 KJV). From our perspective, this is funny. Yet the amount of labor and resources put in to these efforts makes it very clear that those folks were every bit as serious as any of today's space pioneers. The reason is that their perspective was very different from ours. From their point of view, the stars really were just out of reach.

Four hundred years ago, when Galileo peered through his telescope[7] and saw the heavens as they had never been seen before, he declared that the earth was not stationary but that it moved around the sun. This announcement so shook people's long-held fundamental picture of their environment—an earth-centered universe—that it was not until 1992 that Pope John Paul II gave an address in which he admitted that errors had been made by the church in judging Galileo's announcement. The pope did not state what those errors were, but did—some 360 years after Galileo had first been charged with heresy—declare his case closed.

Reality Check

> These are moments when we have to just stop and say, "Hold everything. You can't tell me that this is actually true." The answer is that the truth is unquestionably stranger than fiction. The point, however, is not an indictment of the church. What is amazing—as discussed in the introduction—is how hard it is for us to alter our long-held basic notions of reality.

When Ferdinand Magellan circumnavigated the globe in 1519, proving that the earth was indeed shaped like a ball, people did not just sip their drinks and say, "Well, that's nice." Every idea about how simple ordinary things happened in everyday life had to be thrown out and re-evaluated. This fundamental change in perspective was so shocking that some have still been unable to readjust. An organization known as the "Flat Earth Society" exists to this day. Even now, it maintains that the earth is a flat disk centered at the North Pole and bound along its southern edge by a wall of ice, with the sun, moon, planets, and stars being only a few hundred miles above the earth. The Flat Earth Society was a chief proponent of the idea that the moon landings were faked.

CHAPTER 4

THE PROBLEM OF QUANTUM PHYSICS

> The way I run this thing, you'd think
> I knew something about it.
> —Bugs Bunny

AS WE TOUCHED on in the introduction, when the twentieth century began, two ideas with similar shock value began to shake what many late nineteenth-century scientists had declared to be the true and nearly complete perspective of reality. When Albert Einstein explained relativity, once again every idea explaining the fundamental nature of the universe had to be scrapped. Then, before anyone could begin to piece together anything resembling a tolerable explanation of things, someone threw a riddle into the equations that remains to this very day. In the 1920s, when quantum theory first came to light, what it revealed so astonished the scientific community that it has yet to come up with a logical explanation as to how or why quantum events occur even though these truths have now become an intricate part of our everyday lives.

Quantum physics describes the behavior of the smallest particles known to exist. These particles—quarks, electrons, photons—form atoms and transmit energy between them. Because everything that exists is made of atoms, this, of course, is quite valuable information.

Today, many people consider quantum physics to be the most accurate theory in the history of mankind. There is no question that this is true. The proof is all around us. Quantum physics has brought us television, computers, cell phones, lasers, medical breakthroughs, and a list of technological wonders that grows with each passing day.

What is surprising, however, is that even though we have discovered so much about the quantum world, after almost a century we still have no real understanding of why quantum events occur the way they do. What goes on in the infinitesimal atomic world literally defies common sense and logical analysis. In fact, quantum events are so crazy that they disregard the most basic natural laws that centuries of scientific research have worked to understand and clearly establish. Truths that, for centuries, no one has ever thought to question simply fall apart when you start looking at the way atoms behave. If you are wondering why you haven't heard much about these quantum mysteries after all this time, it is for this very reason. No one really understands them. Scientists themselves have found it impossible to believe what their own research has uncovered.

Since the inception of quantum theory, the greatest minds on our planet have attempted to interpret the behavior of atoms in terms of the physical reality we know and live in. No one has succeeded. That being the case, perhaps it is time to adjust our basic perspective. Of course, as we have discussed, altering such long-held beliefs is incredibly difficult and always meets stiff resistance. As human beings, we find peace and security in the status quo, and the more that new ideas challenge our long-held beliefs, the more anxiety is produced within us. Yet, as we have also seen, such readjustments are anything but new. Men such as Galileo, Newton, Magellan, and Einstein forced dynamic changes in the way people viewed the world and their place in it, regardless of how anyone felt about it at the time. The truth, it seems, has never really had much regard for human feelings or opinions.

CHAPTER 5

THE BIBLE AND "COMMON SENSE"

> The most erroneous stories are those we think we know
> best—and therefore never scrutinize or question.
> —Stephen Jay Gould[8]

DURING THE PAST hundred years, the Bible has fallen far away from popular notions of common sense. The general opinion today seems to be that it is quite "apparent" that the advances of modern science have shown the Bible to be a mostly mythical guide to life. I used the term *apparent* for a reason. You see, at the heart of modern technology lie two theories that, like the Bible, point directly to a reality beyond the natural physical world we think we know. Relativity and quantum physics show us that time and space are not what people have always believed them to be. In the eighty-plus years since the advent of quantum theory, the amount of effort expended to fit it into the "common sense" logical world of cause and effect that most of us assume we inhabit can hardly be imagined. Einstein devoted much of his life to this task. Despite his famous remark that "God does not play dice," neither he nor anyone else since has been able to make any understandable sense out of the quantum world. This has led to a reluctant but gradual acceptance by the scientific community that quantum mechanics really does transcend the laws of physics as we know them and operates in a realm beyond the limits of space and time.

Today, all you have to do is look through the titles in the science section of any major bookstore to realize that there are an ever-increasing number of metaphysical theories that explain our existence—theories that point to a reality beyond what is perceptible to our senses. Of course, the Bible has always maintained that this is true. Yet the very last subjects anyone would ever consider to be addressed by the Bible would be relativity or quantum physics. But if the Bible, relativity, and quantum physics all contain truth, it follows that when each one addresses what is true about the world we live in, they should be in agreement.

Then again, the Bible has been around a lot longer than almost anything else. We see its influence in our culture each time we step out the door, from the church on the corner to the laws down at the county courthouse.[9] That puts it right at the top of the list of things that "we think we know best and therefore never scrutinize or question."

CHAPTER 6

IS IMAGINATION MORE IMPORTANT THAN KNOWLEDGE?

> The true sign of intelligence is not knowledge but imagination.
> —Albert Einstein

AN UNDERSTANDING OF quantum theory requires a different approach. The fact that the latest scientific literature is packed with metaphysical inferences indicates that the traditional modes of analyzing quantum events have hit an intellectual dead end. Of course, suggesting that a metaphysical or "supernatural" perspective could be helpful in better understanding our reality flies in the face of mainstream scientific thinking. However, it may be that a negative gut reaction to such an idea may have more to do with the distrust that has grown on both sides of the religion-science debate than with the actual facts.

Regardless of how vehement a person might be on either side of the issue, there can be no argument that Albert Einstein had a unique ability to see past the established ways of thinking and view things in a way that few people, to this day, ever have. Commenting on this subject, he said, "What is the meaning of human life, or of organic life altogether? To answer this question at all implies a religion. Science without religion is lame: religion without science is blind."

History paints our friend Al[10] as a man unlike any other. However, if we take a minute to look just a little closer, we might find that he was really not as different as one might think. Let me ask you something. Did you ever have a high school teacher who said you didn't know what you were doing and gave you a grade to prove it? Our friend Al sure did. As a child, young Albert did not speak until after the age of two and even for the next couple of years very rarely. His parents, quite honestly, thought he might be mentally retarded. He did so poorly in school that when they asked the headmaster what vocation the boy should take, the gentleman responded, "It doesn't matter, he'll fail no matter what he does." Our friend Al then promptly lived up to that assessment of his abilities.

Apparently not suited to go into his father's electrical manufacturing business after high school, he decided to go to college. That, however, did not mean he could actually find one that would agree with his decision. He failed the college entrance exam … twice. It took an entire year, but our friend Al finally did get accepted, and eventually he graduated. However, that did not mean he could find an employer, including his own college. He put in an application to be a tutor and was turned down. Next, using the logic he was blessed with, he took a step back and applied to be a high school teacher. The result was the same. No one would hire him. Don't forget that one hundred years ago very few people were able to go to college and those who did could have their choice of careers. Anyone, that is, but our friend Al. In fact, for almost two years he couldn't find employment doing *anything*. Finally, a friend pulled some strings and got him a job as an examiner, third class—there was no fourth class—at the Patent Office in Zurich, Switzerland. He took to it right away because the job had built into it plenty of time to goof off.

Now let me ask you another question: Does any of this sound familiar? Isn't it well known that great minds think alike?

It may be that the most amazing quality our friend Al possessed was that he never buckled in the face of constant criticism. He never lost his belief in himself. He never lost his sense of awe and wonder at the world around him. As a young boy, when people told him to stop daydreaming—using his "imagination"—he ignored them. Throughout his life, he retained the eyes of a child. He once said, "The pursuit of truth and beauty is a sphere of activity in which we are permitted to remain children all our lives."

It is interesting to note that when Jesus' disciples asked Him how to find true success, He told them to pursue it in the same manner: as a child (see Matt. 18:1-6; Mark 10:13-16). This was certainly the key to our friend Al's achievements. One famous story relates how he came up with the theory of relativity by imagining himself riding on a light wave around the universe. Later in his life, he said:

> The only real valuable thing is intuition. When I examine myself and my methods of thought, I come to the conclusion that the gift of fantasy has meant more to me than my talent for absorbing positive knowledge.... I am enough of an artist to draw freely upon my imagination. Imagination is more important than knowledge. Knowledge is limited. Imagination encircles the world.

On another occasion, Al observed, "He who can no longer pause to wonder and stand rapt in awe is as good as dead; his eyes are closed." How do children learn? Every-other minute they "pause to wonder and stand rapt in awe." They explore their world and who and what they are through their imaginations. If that process is nurtured and not suppressed or corrupted, the result is always a gift for other human beings as well as personal fulfillment as the child becomes an adult. We use our imaginations to discover ourselves and our world and to become who we were made to be: unique individuals.

When we let our imaginations flow, the result often becomes a picture in our minds. Once we get a picture, it serves as a blueprint or reference for making a dream become a reality. Because this is a prerequisite for any new thing created by humans, it indicates that our friend Al was certainly correct when he said that imagination is more important than knowledge. Imagination is the seat of all human creativity. Without it, there would be nothing new. We might even say that imagination is the most fundamental and potent tool that humankind possesses.

If this is true, then it means that the Bible—if it is what it claims to be—should be in agreement with this concept. Do you remember the story of the tower of Babel? In Genesis 11, the Bible states that everyone on the earth used the same language and the same words. When the people decided to build a tower that could reach to heaven, God put a stop to the project by giving them different languages. Genesis 11:6 states the reason why: "The Lord said, Behold, they are *one* people and they have all one language; and this is only the beginning of what they will do, and *now nothing they have* imagined *they can do will be impossible for them*" (AMP, emphasis added). The King James Version says, "Nothing will be restrained from them, which they have *imagined* to do."

So just like our friend Al, the Bible states that the imagination's possibilities are limitless. But if it is true that before anything can become real it must first exist in the imagination, does contemporary science support this hypothesis? Hopefully, if not running wild, your imagination is now moving at a good steady pace and we can go on.

Reality Check

> If some people give you a hard time for letting your imagination *run* wild, try telling them what our friend Al did. Tell them you are running a "thought experiment." It sounds much better than "daydreaming."

CHAPTER 7

THE MAINFRAME

> It may be that the whole is simple and we are looking at it from the wrong point of view.
> —Henri Bergson[11]

IF YOU GO out to a peaceful place, lie down, and gaze at the sky, you will see the sun, moon, and stars track through the sky in patterns that will soon become familiar. As you relax, you will certainly feel motionless, tranquil, at rest. Observing this, the ancient Greek philosopher Aristotle (384–322 B.C.) came up with the logical conclusion that the earth was a constant, eternal, and steady place that rested at the center of all the movement around it. Today, exactly how right or wrong he was about all this seems inconsequential on the surface. However, its importance is fundamental. The reason is that Aristotle's view became the basic *frame of reference* into which all scientific observations that came afterward were placed. For two thousand years, it was the foundation upon which the fundamental picture of reality was built. Whatever question arose, it was answered within the *framework* of this picture.

Here is an example of what I mean. Let's say that we wanted to meet three miles east and two miles north of 4th and Main Street at 12:30 P.M. to have lunch. In order for us to find each other, we must both use the same frame of reference. That is to say that we both have to agree on what a mile and an hour are. Suppose, just for fun, that between now and 12:30 we both saw every clock running at a different speed and that the earth was unpredictably changing shape in one place and not another. All bets would be off—an hour or a mile might be one thing where I am and quite another where you are. Without an agreed-upon frame of reference, it would literally be impossible for us to effectively function and communicate with one another.

Of course, time does (seem to) pass at the same rate everywhere, and the earth isn't radically changing shape. So Aristotle declared that these two things are constant or the same everywhere. He noted that because space and time—a mile and a minute—are the same for everybody everywhere, the entire universe operated within this basic frame. No matter what problem needed to be solved—whether how to get together for lunch or how the stars move across the sky—the place to start to find the answer was the fact that space and time were the same no matter what was going on or where in the universe it was happening.

Looking at it another way, if you are wondering what might happen today or how something did happen in the past, you will immediately eliminate an infinite number of answers because they cannot fit inside this frame—there is not enough time or space available for these events to occur. Aristotle simply said, "Yes, that's right." As you read on, you will see that not only Aristotle but also a great number of people became famous for putting words to things that everyone—intuitively—already knew.

So those "apparent" facts became the mainframe inside of which all the other aspects of reality had to fit. It worked quite well for about eighteen hundred years and, as a practical matter, still does. However, as history has shown us with all the answers that were absolutely great until—whoops—they were proven wrong, the truth had absolutely no regard even for Aristotle's impeccable logic.

CHAPTER 8

ENTERPRISE

> To seek out new life and new civilizations, to boldly go where no one has gone before....
> —*Star Trek* opening monologue

ARISTOTLE'S VIEW OF the universe began to unravel when two men named Nicolaus Copernicus (1473–1543) and Galileo Galilei (1564–1642) made some observations and decided not to keep them a secret. They announced that the earth was actually not at the center of everything. Now if you think it is hard to change erroneous patterns of thinking that have been passed down through the family for a few generations, just think of what these two guys had to overcome. For nearly two millennia, no one had ever even thought to question Aristotle's view of the earth as at the center of an eternal universe.

Actually, that is not quite correct. Thousands of years before Aristotle was born, the men who wrote the Bible recorded that the earth and the universe were not eternal but had a definite beginning. Unlike Aristotle, the Bible also stated that time and space were not the absolute constants upon which all of reality rested. Galileo and Copernicus began to unravel this old mindset, but it would take another three hundred years before science would find itself in agreement with what the Bible said.

In the year 1492, it was obvious to anyone with two eyes and enough sense to get up and look out the window that the world was flat. Then along came that goofball Christopher Columbus—who, incidentally, stands out as the most completely lost individual in recorded history[12]—who said that he was going to India by heading in the wrong direction. It was only a few years later that the obviously demented Ferdinand Magellan announced that he would head west, keep sailing in the same direction, and eventually end up where he started. All this was good for a few years' worth of great comedy until there turned out to be a problem with these two: they were right. In 1522, eighteen of the 237 original sailors who comprised Magellan's crew showed up in Spain, where they had started their voyage. Magellan himself had been killed in the Philippines.[13] That improbable fact just went and messed up everything. Despite anyone's belief in or fondness for the way things had always been, the voyages of Columbus and Magellan demanded a radical adjustment of one's perspective.

The earth was a big ball and always had been. Yet somehow, nobody was falling off! What did everyone in those days think about that? They thought just what we think about quantum science: "The whole thing makes no sense and is completely crazy." Nevertheless, the truth of the matter continued its policy of having absolutely no regard for popular opinion. This state of affairs left folks with no option but to conclude that some unseen type of glue might be holding everyone on the ball and preventing them from flying off into space in umpteen different directions. What was it? No one had a clue.

Almost two centuries would have to pass before Isaac Newton theorized that the mass of any object attracts other objects with a force in proportion to its size and distance from the other object. Yet even today, scientists are not at all sure how this happens. Many are searching for a particle they believe has to exist to transmit the force of gravity. Some are so sure it must exist that they have even given it a name: "graviton." But so far, all efforts to actually detect such a particle have come up with zilch.

The state of quantum physics today shares this same conundrum. Physicists know atoms and their components behave in certain ways, but they are at a loss to explain the how or why of it. Like a ball that nothing falls off of, the quantum world defies all notions of common sense. Yet as we will see, from a biblical perspective it shares a remarkable synchronicity.

CHAPTER 9

NOTHING HOLDING IT UP?

Hello, ball.

—Ed Norton[14]

AFTER EXPOSING THE fact that the earth was not at the center of everything, Mr. Galileo threw another monkey wrench into obvious reality. He observed that the earth, in addition to being round, was also spinning like a top and moving—along with other planets—around the sun. This made even less sense than a huge ball with people standing on the bottom. Even a blind person can tell if he or she is moving or standing still. If the earth was moving, you would feel it. When you are in motion—run, ride a horse—you create a breeze. If the earth was really spinning at hundreds of miles per hour, how could there be a calm day? It would play havoc with the ocean, not to mention everyday life. Besides, people would get dizzy.[15] Obviously, the whole thing was ludicrous—just like quantum mechanics is to us.

Unfortunately, if all of this was hard for everyday people to handle, it was even more difficult for the scholars and theologians of the day. At this point the church leaned on its own understanding[16] and made a critical mistake that has haunted science and religion to this day. It sought to silence these discoveries along with the men who made them. And in those days, it had the power to do it!

There was no division of church and state at that time. There never had been. The Roman emperors had made themselves gods. When Emperor Constantine declared Christianity to be a state religion in the year 313, this unity did not change. He conveniently appointed himself Christ's supreme representative on earth at the same time. As Hal Lindsey notes in his book, *The Road to Holocaust*, it did not take long for this new and powerful church to proclaim *that they were the inheritors of the kingdom promised to Israel and therefore must take ultimate authority over the political powers of the world.*[17] Essentially, this new all-powerful, all-encompassing church believed that it had inherited David's throne and that all of the Bible's prophecies concerning Israel's future referred to itself.

After the Roman Empire fell, the only stable organization that remained in western civilization was the church. During the "Dark Ages" that followed, this furthered the belief that the church leaders were God's special representatives, chosen by Him to hold their offices, interpret Scripture—define morality—and lead the masses. The kings and queens who came to power out of the ashes of Rome continued, like their Roman predecessors, to believe that they were beings of a higher order chosen by God to lead and rule. The common belief throughout all levels of society was that kings had a "divine right" to lead and therefore—along with the clergy—had been given special powers and privileges by God. They believed, as most everyone on earth always had, that God ordained the levels of society. Just as bloodlines determined who the next king would be, ancestry governed all levels of society. Who you were was unchangeable; God determined it at your at birth. The higher your position, the more favored you were by God and the more authority God had given you over others. Also, the church leaders were the only ones who had copies of the Scriptures. This most vital information in the control of a select few gave those who possessed it tremendous power and influence. People actually believed that a priest had the ability to decide whether a person went to heaven or hell.

In the year 1439, however, something happened to plant seeds of change. It was then that a new information highway was invented: the printing press. In 1534, that upstart Martin Luther went and translated the Bible into a language (German) that the people in his country

(Germany) could actually understand. This death-defying feat was copied in other countries.[18]

When Luther began to read the Bible for himself instead of having someone interpret it for him, he was *astonished* to find it said that faith in what Jesus had done for us, not anything we did or anyone else—including a priest—did or decided, was the basis for salvation. Not only that, but Luther also discovered that the Bible said Jesus died so that each of us could actually have our own personal direct relationship with God (see John 14:6, 16:13). For some reason—what a surprise—those in charge had not passed this around. This was *completely contrary* to more than a thousand years of unquestioned church practice. Jesus' sacrifice had actually abolished the requirement of an intermediary (a priest) between God and man. He had made us *all* priests (see 1 Peter 2:5). He had made us *all* equal. He had made us *all* children of the same Father: God.

Needless to say, there was more than just stiff resistance to the spreading of this information. But as it did, this *protest* fractured the church into the all of the *Protest*ant denominations we have today.[19]

The invention of the printing press meant that slowly but surely, everyday people were gaining the ability to obtain the information contained in the Bible and interpret it for themselves. As a result, a belief grew among the populace that, contrary to centuries of tradition, all men were created equal. That led in the 1700s—for the first time in history—to the formation of a country based upon that premise.[20]

However, at the time Copernicus and Galileo were looking through their telescopes, the accepted belief was that a priest had the ability to decide whether a person went to heaven or hell. Ole Nick, being the brilliant thinker he was, decided to beat the system and not publish most of his findings until after his death. That way no priest on earth could excommunicate him—keep him out of heaven. He was already there.

Reality Check

> I just have to stop a minute and take my hat off in admiration to this the kind of thinking. It's inspiring.

It was the rediscovery of Greek civilization that spawned the Renaissance. A thirteenth-century cleric by the name of Thomas Aquinas is generally credited with beginning this "reawakening" when he discovered Aristotle. He then began a synthesis of Aristotle's explanation of the universe with the Bible. In Aristotle's time, his ideas were freely debated. However, once Aquinas was finished, that was no longer the case. The Bible was not debatable, and because it was now believed that Aristotle's description of the universe was the same as the Bible's, obviously, Aristotle's view could not be questioned either.

This bugged Galileo. He stubbornly held on to the idea that if you could see with your own two eyes that Aristotle was incorrect about something, it might actually be so. His insistence on this was widely regarded as a bad career move. Aristotle said that heavier things fell faster because their weight meant they had a greater desire for rest at the center of the universe—which was the earth's center. Galileo said that a feather falls slower because of wind resistance. He is then thought to have climbed the Tower of Pisa in broad daylight and have dropped a lead ball and a piece of wood off of it only to actually see them hit the ground at the same time. To no one's surprise, the church of the Inquisition informed him that for his own good—as well as everyone else's—it had decided to permanently remove his ideas from further circulation. To achieve this, they developed a straightforward strategy: a sufficiently sharp edge would be requisitioned to carve out an adequate distance between the mind generating these thoughts and the body supporting it.

Faced with this impediment to continued cognition, Galileo decided his brain would be of little use to anyone—especially himself—if it could not retain close proximity to its supporting extremities. He therefore declared that he didn't see what he saw. The church immediately declared this to be evidence of Mr. G's sanity. In recognition of this, it permitted his mind to remain in contact with the rest of himself as long as he went someplace where he wouldn't bother people with preposterous and obviously crazy notions like the earth was moving and just hanging in space with nothing holding it up.

That was less than four hundred years ago. However, it actually was not the first time someone had made such an outrageous claim. About four thousand years ago, the author of the book of Job stated that God stretches out the north over empty space and hangs the earth on nothing (see Job 26:7).

The church's unwillingness to listen to Mr. Galileo—let alone its own Scripture—can only logically be attributed to the all-encompassing fear of losing the power and influence it had accumulated over the centuries. The result was a spiritual and intellectual split that exists to this day. Seekers of truth became divided into two separate camps: discerners of physical truth—scientists—and seekers of spiritual truth—clergy. No such division existed before this time.

Aquinas had done nothing new in creating his synthesis of Aristotle's explanation of the universe and the Bible. The Greeks and many societies before them had always viewed the spiritual and physical as co-dependent, with each unable to be understood without contemplation of the other. However, when Galileo and Copernicus tried to talk about what they had seen through their telescopes, that idea flew out the window and hasn't been seen since. The prejudices and skepticism that emerged between science and religion in the 1600s continued to grow through the centuries that followed to the point that today the common belief is that the study of one is detrimental to understanding the other. Today, the scientific community seems to assume that religious belief is fantasy and inherently incompatible with known physical reality. Teachers of religion also seem to assume that mainstream scientific knowledge is inherently incompatible with spiritual truth. Both disciplines now see the other as an obstacle to the discovery of the truth.

But why? One of the things that history has shown us is that the truth has absolutely no regard for whomever it is that pursues it. Whether evolutionist or evangelical, the earth will spin, water will remain two parts hydrogen and one part oxygen, spouses will continue to experience moments of utter and total bewilderment no matter how long they've been married, and where we came from won't change no matter who accuses whom of blasphemy, religious intolerance, or just plain stupidity.

Today, radio telescopes scan the heavens twenty-four hours a day for signs of extra-terrestrial intelligent life. At the same time, the latest and most compelling scientific explanations of the reality we inhabit—which we will look at later—tell us that there are more dimensions than the four (length, width, depth, time[21]) that we call home. One book that's been around for a while, the Bible, also states that there is extra-terrestrial intelligent life and that it exists in dimensions outside these four walls.

Reality Check

> Imagine that you are sitting across a table from someone, and that in the middle of the table is a piece of paper with the number 6 written on it. The next day, when that person asks you what was on the paper, you say there was a 6. He says, "You are crazy, it was a 9." Unless the other person is standing in your shoes, he or she is not going to have the same view of things that you have. It is not humanly possible for one person to see everything the way another person does. Just ask anyone who has been married for more than a couple of days. Could it be that science and religion are sitting at a table across from each other, arguing about whether a 6 or a 9 is written in the middle?

CHAPTER 10

CONSTANT MOTION

> When you are courting a nice girl, an hour seems like a second.
> When you sit on a red-hot cinder, a second seems like an
> hour. That's relativity.
> —Albert Einstein

OKAY, COLUMBUS DIDN'T sail off the edge of the earth, and Magellan's crew came back. So maybe the world is round? But, obviously, it is not moving. Everyone knows the difference between moving and standing still. Galileo said that is not necessarily so. He said that if you are moving at a constant rate of speed, the world does not behave any differently than it does when you are not moving at all. Here in the twenty-first century, where we fly and ride in cars every day, this is a lot easier to get across than it was in 1630.

If you are flying in a jet plane at a constant speed in calm air with the windows closed, it can be hard to sense movement and virtually impossible to get a sense of how fast you are traveling. If you drop something or pour a drink on a plane, things happen just as they would if you were doing them at home. It is only when you are in the process of taking off or landing that you can get a sense of how fast you are really moving, and the world doesn't cooperate as it should—the drink spills.

The principle is that the laws of physics are the same when a person is standing still or moving at a constant speed. Galileo explained it by saying that if you were inside a ship's cabin with all the windows closed on a calm sea, you could not tell if you were anchored or sailing because there would be no difference between the way things behaved in either case. If you dropped something or watched a fly buzz around the room, the behavior would be exactly the same. From your perspective inside the boat, you would see yourself at rest whether you were actually anchored or moving at a constant speed.

Although, as I said, this is not very hard for us to grasp, it was this understanding that eventually overturned Aristotle's basic premise of a universe of constant space and time with an earth at rest at the center of it all. Remember our discussion in chapter seven of the fundamental impact this idea had on the discoveries that followed? For two thousand years Aristotle's view of things seemed as obvious as the nose on your face and had been the basic *frame of reference* into which all scientific observations that came later were placed. It had been the foundation upon which the fundamental picture of reality had been built. Whatever the question, the answer existed inside the *frame*work of that picture.

Galileo pulled the frame apart, and no one was happy. People tried to shut him up and put it back together before anyone noticed. The truth, however, continued its annoying habit of popping up at the most inopportune moments despite overwhelming sentiment that it go away and stop bothering people.

Not only was this principle that the laws of physics are the same when a person is standing still or moving at a constant speed a fundamental concept of Galileo's view of our world, it was also the key that unlocked Albert Einstein's explanation of the universe. The name "relativity" became popular when people began to realize the full ramifications of this principle. If these same ramifications don't immediately come to mind, don't worry. Nearly three hundred years passed between Galileo and the time the spark went off between our friend Al's ears—and, even then, few believed him.

CHAPTER 11

OUR OWN POINT OF VIEW

Who are you going to believe, me or your own eyes?
—Groucho Marx

SO WE HAVE the principle shown to us by Galileo that the laws of physics are the same regardless of whether one is standing still or moving at a constant rate of speed. After a few centuries, almost everybody had gotten used to this idea until our friend Al came along and once again turned reality upside down and sideways. All he did was ask a simple question: "What happens if two people are looking at the same event and one is moving while the other is not?" That seems easy enough. Who knows? Maybe it is.

Let's take a closer look at this by going from Galileo's boat to the beach (see figure 2 below). Here we see Ethel sitting still on the sand watching a boat go by. On the boat is Fred, who is bouncing a ball on the deck. The boat is traveling on a calm sea at a constant speed. So from Fred's point of view, things are behaving just as they would if he was not moving at all. The ball is going straight down and straight back up into his hand.

However, this is not what Ethel sees from the shore. Ethel sees the ball move in a pattern that looks like a V. Between the time the ball left Fred's hand and hit the deck of the ship, Fred, the ball, and the entire boat moved forward. The boat also moved farther as the ball went back up. This means that for Ethel, instead of seeing the ball take a straight path down and straight back up the way it came down, the path shifts a bit. As she tracks the ball with the sky as a backdrop, it moves in a pattern like a V. Because the boat was moving with Fred on it, she saw the ball take a longer path than Fred did. The path each person saw the ball take was *relative* to his or her point of view—his or her *perspective*.

Later, when the two discuss what they have seen, Fred says that the ball went straight down and up. Ethel says, "No way. It definitely moved in a V-pattern." Fred, who has seen every Marx Brothers movie, indicates that she did not really see what she just saw. Ethel, much like Galileo, sticks to her guns. Eventually, both hire lawyers to prove that the other is crazy, and the debate—like the one between science and religion—goes round and round.

It is not too difficult to understand two people having a different perspective of the same event. The hard part, as we will shortly see, is grasping the impact that this simple difference has on our perception of the reality that we think we know.

CHAPTER 12

GRAVITY

I know this defies the law of gravity, but I never studied law.
—Bugs Bunny

TODAY, WE HAVE what most of us would call commonsense ideas of how the world functions. We inherited these concepts through previous generations who had accepted them for a very good reason: they worked and proved to be repeatedly reliable. So quite naturally, these ideas became commonplace and were taken for granted. But as we have seen, commonsense things that we don't even think to question now were anything but obvious in times past. With that in mind, let's go back to 1687.

There you are, a model citizen minding your own business, when along comes this nut who says that what causes a glass to break when it falls is the same thing that caused the tide to come in this morning. You run a quick assessment of his mental condition, but before you can give him the benefit of the doubt, he goes on to tell you that it is also the same thing that causes the moon to go around the earth. When you gently ask him his name because you know some nice folks (in white coats) who like to help people who think watching apples fall off trees will benefit mankind, he answers, "Isaac." He tells you, "I'm thinking of calling it either tribble or gravity." To amuse him, you vote for "gravity," and the rest is history. One becomes the name of a fundamental force, and the other becomes the name of a furry creature in a Star Trek episode.

Isaac Newton, like Magellan and Galileo, fundamentally changed the way human beings viewed their relationship to their surroundings. After Sir Isaac published his findings in 1688, humankind saw themselves in a much more important position in the cosmic scheme of things. Instead of being subjected to the whims of mysterious (and therefore fearful) forces, Newton revealed that we live in a universe that actually behaves quite rationally. The planets are not being held up or pushed around by angels. They, along with the tide and the glass that fell to the floor, are all being governed by the same force that keeps everyone from floating to the ceiling.

This force of gravity turned what had been unfathomable mysteries into events that could be easily understood by the rules of cause and effect. Knowing these rules meant that a person no longer needed the help of mysticism or superstition to understand life. Logic and mathematics could do the job. Nature was not something to be feared. Not only could it be understood; it could also be used. Suddenly, the future became one of hope and boundless possibilities. A universe based on cause and effect meant that if a person kept on linking the two, whatever was unknown could eventually be known. Famine and disease could be conquered. Like a rational and orderly mechanism, the universe could be taken apart and understood. Not only did this revolutionize the way people viewed their surroundings, but it also fundamentally changed the way they saw themselves. Instead of being helpless victims of their environment, mankind could now be master of it.

The impact that this change in perspective had on society cannot be overemphasized. People now saw the universe as a mechanism that God had set in motion and given those He created in His image the ability to understand, manipulate, and control. Science, as we know it today, was launched by Isaac Newton.

One of the effects this new perspective had was that, as people began to view themselves as more important in the universal scheme of things, they began to lower the importance they placed on God in their personal lives. It was not that they no longer believed in God as the Creator, they just viewed His role—and, therefore, their own role in the cosmos—differently. God was now seen as one Who had set a mechanical universe in motion and then left it to His creation (us) to manage. He had established the natural laws by which the universe works, and we humans, being created in His image, had been given the ability to understand and use those laws to solve problems.

Because of this new perspective, the concept of a personal God began to be relegated to the category of outmoded superstitious beliefs. God was no longer viewed as a personal savior to Whom you should run every time there is a problem. In fact, people who called on God to be involved in the daily aspects of their lives were seen—and are still seen by many—as "unenlightened," fearfully refusing to let go of the myths of the past. Although praying to God was not a bad thing, just sitting around doing that was considered a cop-out. It was incumbent upon humans to use the grey matter that God had given them to learn, discover, and "enlighten" their minds. God helped those who helped themselves.[22]

One could use this new "enlightened" perspective to not only unlock the mysteries of the physical universe, but also the mysteries of human behavior. Using the laws of cause and effect, one could now predict that a person's behavior (the effect) occurred as a result of a prior experience and/or a genetic predisposition to do that behavior (the cause). The social sciences and the fields of psychiatry and psychology are the result of this basic idea. People were now believed to be the product of how their past experiences affected the genetic blueprint they had inherited from their parents. Today society wrestles with the logical consequence of this concept. If what a person does is ultimately predetermined, then that person cannot be held responsible for his or her improper actions and bad behavior.

Reality Check

> Regardless of how true it may or may not be that our behavior is predetermined by conditions over which we have no control, history tells us that the success of any group (family, neighborhood, or society) is directly dependent on the degree to which each individual, irrespective of personal circumstances, is willing be responsible for how he or she treats others. The essence of Jesus' message in word and deed was to love our neighbor and do unto others as we would have them do unto us. We do this not because of what has happened to us or who we are but because of who *they* are in relation to God: His children.[23]

The laws Mr. Newton gave the world were used to build the machines that powered the Industrial Revolution. These laws work so well that one could go so far as to calculate exactly how much highly combustible liquid propellant it would take to stick a man in a can—a space capsule—shoot him free of the earth's gravity and predict—within a few thousand miles—where he might come down. Admittedly, that is a loose description of NASA's Mercury Space Program of the early 1960s (though maybe not so much from the astronaut's point of view). The point is that Newton's laws have worked so well for so long that today we take them for granted.

CHAPTER 13

"UNIVERSAL CONSTANTS"?

> Nothing is more dangerous than a dogmatic worldview—
> nothing more constraining, more blinding to innovation.
> —Stephen Jay Gould

ISAAC NEWTON'S PERSPECTIVE of space and time, like Aristotle's before him, became the mainframe inside of which all of nature's laws worked. No matter where in the universe a person happened to be, those laws did not change.

These types of basic, unquestioned concepts that everyone agrees upon are known as "universal constants." As was stated in chapter 7, they provide the "mainframe" inside of which each of us constructs our view of the world we live in. Everything we think about doing from the moment we wake up fits within this environment of space and time. As we form an idea of whatever it is that is going to happen (or ever did happen), we paint that picture inside of that frame. This is a good thing because it allows us to coordinate our actions with others. Because a ton, a gallon, a mile, an hour, and so forth are the same for all of us, we can calculate and communicate all sorts of ideas to each other and accomplish our goals. In fact, these universal constants are so basic to our existence that we don't even think about them. No one needs to explain how long an hour is; it is so ingrained in our brains that many of us wake up a minute before the alarm goes off on a regular basis.

So just for fun, let's assume that a few apparently "starved-for-attention" people come along and, with a straight face, announce that distance and time are different for different people. An hour for one person is really not the same as an hour for another, and, for that matter, neither is a mile.

Reality Check

> Yes, we would ignore them and ask someone to drive them home.

Okay … now imagine what would happen if you discovered that some of these people actually started pointing this out more than one hundred years ago.

Reality Check

> I guess you would say that one universal constant is that there is no time limit on stupidity.

Now what would you think if I told you that one of these people went by the name of Albert Einstein?

Reality Check

> Uh-oh …

Could it be that our perception of reality became so ingrained in us that it has taken nearly a century for us (and a good many scientists) to begin to open our minds to the real ramifications of relativity and quantum physics? Could it be that these truths really do show us that Sir Isaac's laws are not the foundational principles that we, for so long, have thought are governing the universe?

CHAPTER 14

WHAT'S REALLY OUT THERE?

> When it comes to humility, I'm the greatest.
> —Bullwinkle the Moose

FIVE HUNDRED YEARS ago, no one had any idea that there would ever be a need to change the way they looked at the world they lived in. But Columbus made his voyage, Magellan's crew completed their circumnavigation of the globe, and, like it or not, everything people thought they knew about what they saw outside their window had to be thrown out. It was back to the drawing board. A century later, things had hardly settled down when Galileo looked through his telescope and announced that the earth was in motion and not at the center of the universe. These discoveries meant that the fundamental picture people had of the most common everyday occurrences, such as why things fall down and why day and night occurred, had to be reconstructed from the ground up.

It wasn't until about eighty years after Galileo's observations that Isaac Newton began to make some kind of sense of it all by introducing the force of gravity into the picture. Actually, what Sir Isaac said made more than some sense. He sparked a scientific revolution that became an industrial revolution. Finally, by the time the twentieth century began, the general picture of reality had been entirely rebuilt. The constant, predictable universe of cause and effect that Newton uncovered had been studied, measured, researched, and verified to the point that it was as taken for granted as the sun coming up in the morning. By the early 1900s, nearly all of the questions and apparent discrepancies that had initially been raised by Newton's observations had finally been resolved. Except for the occasional clash of incredibly overblown egos, all was peaceful and well settled in the scientific world. Summing up the situation, Lord Kelvin, one of the leading scientists of the time, declared in his distinctly British accent, "There is nothing new to be discovered in physics now. All that remains is more and more precise measurement."

Reality Check

> Whoops.

Unwilling to let sleeping dogs lie, in the 1920s a man by the name of Edwin Hubble went and did it again. What did he do, you ask? Just like Copernicus and Galileo, it seems that he just couldn't resist looking through a telescope and telling people what he saw. Instantly, not only did the past three hundred years of cosmic research crumble into rubble, but so did every idea that man had ever even dared to dream about the universe. Just like Galileo before him, Hubble messed with everyone's—even Einstein's—understanding of the cosmos.

Looking through a new one-hundred-inch telescope (the world's largest at that time) at Mount Wilson, California, Hubble discovered that what had always appeared to be just blurs of light inside our own Milky Way galaxy were actually separate distinct groups of billions of stars (other galaxies) just like our own Milky Way at incredible distances. Until then, people had presumed that everything they saw in our own galaxy was the extent of the entire universe.

Reality Check

> This distinctly human tradition of presuming that whatever we can see is all that actually exists has continued uninterrupted since the earliest Chinese astronomers built towers to reach the stars. History, however, tells us that the imaginations of previous generations have been no match for the actual truth. Is there any reason for us to think we are any different?

Current estimates are that the Milky Way is one of close to two hundred billion galaxies.[24] This, of course, was an amazing discovery. However, it was something Hubble discovered about these galaxies that once again caused the prevailing scientific perspective—that had taken all of human history to craft—to be scrapped and sent everybody back to "square one." It seems that wherever Hubble looked, he noticed that all the galaxies are traveling away from each other at breakneck speed. Pondering all this, scientists concluded that the farther back you went in time, the closer the galaxies were to each other. Rolling the film backward, current estimates are that approximately 13.7 billion years ago everything in existence was in a very small single place the size of a pea (called a "singularity") of almost infinite density and mass, until it went "Bang"—a very "Big Bang."

The latest thing to come off the drawing board is this. One would think that, like any explosion, the rate at which the universe is expanding would slow down as the initial energy from the Big Bang dissipates through time and space. Recent observations, however, have revealed that, instead of slowing down, the speed that the galaxies are traveling away from each other is actually *increasing*. In light of what we know about energy and gravity, this is impossible.

There is, however, an even more fundamental mystery concerning gravity. As scientists surveyed the vastness of the universe, it became clear that not only was there not enough gravity to cause galaxies to move the way they do, but there was also not nearly enough visible mass in the universe to have allowed gravity to pull two atoms together, let alone form a star or planet. The universe is simply way too big and the amount of matter we know of way too small. However, since we exist, there has to be another explanation.

Reality Check

> Here I would like to point out that even though you are not a theoretical physicist, once you became aware of the facts, it took you exactly the same amount of time to come to this conclusion as it did the most educated people on the planet. The only difference is that instead of producing two hundred pages of rhetoric you went, "Like duh," and, in essence, said the very same thing.

So from our perspective here in the second decade of the twenty-first century, the only logical conclusion is that there must be a whole lot more matter and energy out there that we just can't see.

Now when I say "a whole lot," that is no exaggeration. According to the laws of physics as we understand them, the only way to account for the fact that we do exist and the stars, planets, and galaxies did form as they have is to assume—a dangerous but unavoidable endeavor—that about 96 percent of the entire universe consists of what physicists have called "dark matter" and "dark energy."

The reason they are called "dark" is very simple. No one has ever *seen* or directly *detected* either one.

"Assuming," then, that 96 percent of the universe does actually consist of matter and energy that no one has ever seen, scientists believe that of the remaining 4 percent of matter in the universe—that we actually *do* know about—approximately 3.7 percent of it consists of gas out in space between the stars and galaxies. This means that according to current scientific estimates, all the galaxies, stars, planets, people, insects, stuff at 7-Eleven—*everything*—comprises less than one half of one percent (0.4 percent) of all the matter in the entire universe.

So instead of a constant predictable earth at the center of everything with stars that a tower might reach, or even a motionless sun with planets flowing in orbit around it, we now know what you had already surmised quite some time ago. It's a really big universe and no one has any real idea at all about what's going on out there.[25]

CHAPTER 15

THE UNCERTAINTY PRINCIPLE

> If you thought that science was certain—well,
> that is just an error on your part.
> —Richard Feynman[26]

ON THE OPPOSITE end of the cosmic panorama is the study of the very small. One hundred years ago, Mr. Einstein and others experimentally confirmed the existence of extremely small basic building blocks of all matter called "atoms." And like Hubble, instead of helping out dear Lord Kelvin, they opened a can of worms that has become even more troublesome than any of Hubble's observations.

In fact, Einstein himself spent a good portion of his life trying to put the worms back into the can. However, reality threw him—and everyone else—such a curve that only a very few people were able to even get a piece—small understanding—of what he and his contemporaries uncovered. To this day, there are many who still find it impossible to believe what the results of more than eighty years worth of scientific experimentation and research have repeatedly confirmed. To put it another way, the truth has had absolutely no respect for what the people seeking after it believed it would, should, or even could be. In fact, it really was too much of a stretch for even the most fertile scientific imaginations.

It was Isaac Newton who first explained how one could know exactly when and where two objects—billiard balls, for example—would be at any time in the future by measuring their current location and velocity. If you measured both, you could predict when the balls would collide, where they would go afterward, and when they would get there. It follows, then, that one should be able to predict the behavior of smaller objects—such as atoms—in the same way.

Once the existence of atoms was confirmed, scientists quickly set about constructing devices to detect and measure how they moved. Their motivation was simple. While learning how to manipulate pool balls might earn you a little money, learning how to manipulate atoms might unlock the secrets of the universe. Unlike eight balls, atoms are the fundamental components of everything, including you and me. So as our friend Al and his contemporaries uncovered this atomic world, the possibilities seemed limitless.

What happened next, however, so shocked the scientific world that only now, almost a century later, has what they found finally been accepted by more than a handful of individuals as an actual fact. Eighty years of testing, retesting, calculating, and recalculating have not been able to alter the findings. In fact, all attempts to disprove what was first uncovered more than eighty years ago have, instead, inescapably confirmed it.

The comfortable and "certain" picture of reality that Newton's laws had given the world began to unravel in 1927 with a phenomenon that Werner Heisenberg called the "Uncertainty Principle."[27] What Heisenberg said was that he—or anyone else, for that matter—could not be certain of *the exact speed* and *location of atoms*.

Reality Check

> Huh?

You can know the exact speed of an atom, and you can know the exact location of an atom. What you cannot know is *both* the exact speed *and* the exact location of a single atom. Now your first reaction to this might be—as mine was—that the technology to determine both of these things was lacking during Heisenberg's time or is still lacking today. But that is not what the Uncertainty Principle is about. What it means is that it is not possible for anyone to ever know these two things about a single atom at the same time. You can know one or the other, but no matter what tools science may develop to see into the atomic world today or in the future, both *the exact location and exact speed of atomic particles can never be known.*

Reality Check

> This is a classic instance in which the words themselves are a common part of everyday language. The difficult part is believing them. Last time anyone checked, pizza, cars, and everything else that exists are made entirely out of atoms. How could you deliver a pizza or eat it if you didn't know both the speed and the location of the atoms you and it are made of? Yet, hard as it may seem to believe, once you break a pizza or a person into pieces the size of atoms, all bets are off. For nearly eighty years, *every* attempt to individually measure both the speed and location of atoms and their components has only served to repeatedly confirm Mr. Heisenberg's uncertainty principle.

Thus, if you know the exact speed of an atomic particle, you will not be able to have any idea at all of where it is. As far as you will be able to determine, it could be located anywhere in the universe. Conversely, if you know a particle's exact location, as far as you will be able to measure, its speed could be anywhere from less than one inch per century to 186,282 miles per second.[28]

This does not mean that the speed and the location of an atom are not related; they are. The amount you know about one will have a direct bearing on how much you can know about the other. For example, if you are 70 percent certain of an atom's location, you can be 30 percent certain of its speed. Likewise, if you are 70 percent certain of its speed, you can be 30 percent certain of its location. This correlation is an important principle, because it is on these probabilities that quantum mechanics rests. A television picture tube throws billions of atomic particles—electrons—at the screen. It is impossible to know if any single one of them will actually reach the screen, but it is possible to know that a certain percentage—say 70 percent—will get there and give you something to watch.

When our friend Al said that "God does not play dice," this is what he was referring to. When you roll the dice, you can never be certain when a seven will come up. However, over a large number of rolls, you can approximate the percentage of times that one number will come up as opposed to another. The same is true in predicting both the location and velocity of an atomic particle. As a result, quantum science moves ahead and casino owners get very rich.

Reality Check

> If anything you have just read sounds crazy to you, then congratulations. *It is crazy.* However, your ability to discern this fundamental fact puts you right up there with a good many scientists. Why, you might ask? Because that is exactly the same conclusion they have reached. Science knows that quantum mechanics works but, quite simply, no one is sure *how or why*.

Admittedly, I have poked fun at the scientific profession. However, without a doubt, their amazing insights and discoveries during the last century have transformed the planet. I am thankful for the incredible ingenuity, genius, and just plain hard work of the scientific community. They have made our lives longer, healthier, and happier. However, even though we have made astounding technical advances, of what can we actually be certain? As you already know—or have at least strongly suspected—if I may paraphrase Richard Feynman's observation at the beginning of the chapter, *about the only thing certain is that nothing is certain. If you think you have something figured out, that's just an error on your part.*

So if one of the greatest physicists of all time concluded that nothing is certain and that whatever people think they have figured out is going to turn out to be wrong, where does that put us? That's correct: right there in the same boat with him. In fact, it might put us slightly ahead. Apparently, Mr. Feynman did quite a bit of research before he came to this conclusion. We knew it right from the start.

CHAPTER 16

LIGHT SEED

> Many people do not like the idea that time had a beginning
> probably because it smacks of divine intervention.
> —Stephen Hawking[29]

IMAGINE A SMALL apple seed in the palm of your hand. How much food do you have? The answer is enough to feed everyone on earth and everyone who will come after. Inside that seed is a tree with hundreds of apples and thousands of seeds, each with the same potential as the one in your hand. It seems that modern cosmology has now given us the same picture of our entire universe.

Sometime around 1400 B.C., Moses, who most theologians believe is the author of Genesis, wrote that the universe did not always exist. It had a beginning. Moses said that God had created everything, including the stars, sun, and moon. This was in stark contrast to the popular belief at the time, which was that these celestial bodies were eternal manifestations of individual gods.

More than a thousand years after Moses left the planet, Aristotle said that the universe had always existed and that it always would. This was the unchallenged and accepted scientific view—shared by many theologians—for so long that by the mid-twentieth century, an eternal universe was simply accepted as fact. Even Einstein subscribed to it. He did, however, reveal that he had an open mind on the subject when he said, "Only two things are infinite, the universe and human stupidity, but I'm not sure about the former."

However, after Edwin Hubble's observations in the late 1920s, the fact that the universe was expanding became undeniable. This meant that if you rolled the film backward about 13.7 billion years, everything in existence was squeezed into a pellet the size of a tiny seed. Where did the seed come from? No one knows. Before that, space and time did not exist, which puts the answer to that question beyond the realm of physical reality—physics. So although it took a few thousand years, in the twentieth century science came into agreement with the Bible by stating that the universe did have a definite beginning—and that when it did begin, it must have made a lot of noise.[30]

However, after centuries of disagreement, name-calling, and distrust, most of the world is still under the impression that the biblical and scientific views of creation are at odds. In fact, they are not. In chapter three, I mentioned that in 1992, the Vatican formally admitted the mistakes it had made in judging Galileo. It should be pointed out that prior to that, Pope Pius XII (1939–1958) had privately expressed regret for that error. In 1951, he declared that science had finally realized what the Bible had always maintained. The Bible says, "In the beginning God created the heavens and the earth" (Gen. 1:1). Both the Bible and the Big Bang Theory say that there was nothing until—*bang!*—suddenly, there was everything. It is just that the Big Bang Theory doesn't speculate as to what caused that event.

It is important to understand that the Big Bang Theory does not state that there was a vast empty space and then an explosion filled up the emptiness with stars and planets. What the theory says is that before the explosion there was *nothing*. And when I say nothing, I mean erase everything, including space itself. The theory says that there was no space. It says that there was no time. It says there was no existence. It's hard, if not impossible, to imagine absolutely nothing, yet both the Bible and the Big Bang Theory agree that there was absolutely nothing until matter, energy, space, and time itself burst into existence.

For example, the Bible states:

- He created the worlds and the reaches of space and the ages of time [He made, produced, built, operated, and arranged them in order] (Heb. 1:2 AMP).
- By faith we understand that the worlds were prepared by the Word of God, so that what is seen was not made out of things which are visible (Heb. 11:3).
- Abraham believed in the God Who brings the dead back to life and Who creates new things out of nothing (Rom. 4:17 NLT).

The Big Bang Theory also says that what is seen was made out of things that are not visible. It says that there was nothing, and then suddenly, out of non-existence, a singular speck of light came into existence. This light energy was so intense that out of it came every atom that currently exists in every thing—be it a distant star or your lunch.

Genesis 1:3 states that the very first physical thing that existed was light. Note that this does not refer to the creation of the stars or the sun—that does not occur until Genesis 1:16-18.[31] The Big Bang Theory actually describes this same process. It states that there was an initial explosion of light energy. Out of this primordial light came the building blocks for simple atoms, such as hydrogen and helium. The theory says that some two hundred million years later, under the influence of gravity, these initial atoms began to clump together to form the first stars. However, the more complex elements, such as oxygen, carbon, copper, and lead, were not created until after these stars formed and then burned out in supernovae explosions. Contemporary scientific theory states that hydrogen and helium—and possibly a small amount of lithium—were the only elements produced by the Big Bang. All of the heavier elements were forged inside dying stars. That means that current scientific theory states that exploding stars created 90 percent of the atoms in our bodies. It also means that today's science says that hydrogen and helium eventually evolved into something that could write and read this book.

Reality Check

> On the surface, that may seem like a sarcastic remark. It's not. Many modern scientists take this approach. If you look through the science section in your local bookstore, you will see titles referring to the "Conscious" or "Self-Aware" Universe. Leaps of faith are required regardless of what direction one approaches in trying to understand our existence.

Just like the Big Bang Theory, Genesis says that light was the first thing to exist and that out of this primordial light the stars were later created. In the years to come, scientists will speculate as to where this singularity—light seed—came from. Yet no one living inside space and time can answer such a question. How can you ask where something came from if there was no place for it to exist in the first place? How can you ask what happened "before" if time itself did not exist?

One thing is certain: Such obstacles won't stop us from trying. As a result, we are seeing ever-increasing speculations about other dimensions and/or multiple universes coming from theoretical physicists. This, again, is completely consistent with the Bible since it emphatically states that the source of this universe—God—exists outside the four dimensions of space and time we call home.

CHAPTER 17

A MATTER OF TIME

> One cannot really argue with a mathematical theorem, so it is now generally accepted that the universe must have a beginning.
> —Stephen Hawking[32]

ACCORDING TO THE Big Bang Theory, everything blasted into existence from nothing. Gravity gradually gathered the initial hydrogen and helium together to form stars, which then gave light to the universe. They accomplished this through the process of nuclear fusion—changing hydrogen into helium.[33] Once a star's supply of hydrogen runs out, it dies. In the case of one particular star we might be curious about—our sun—that event will occur in about five billion years. For the most part, when a star runs out of hydrogen, it burns out in an unspectacular way. However, once every fifty years or so in a galaxy of billions of stars, one of these dying stars collapses with such force that it produces a supernova explosion.[34] As we saw in the last chapter, it was these supernova explosions that produced all the other elements in the universe.

In total, these ninety naturally occurring elements are the building blocks of you, me, and everything else in existence. Current cosmological theory says that our own sun and solar system came into existence about five billion years ago. The Theory of Evolution says that some four billion years after that, on this particular clump of elements we call earth, there was a "little bang." Somehow, somewhere, on some microscopic piece of this planet, something miraculous happened—(bang)—life suddenly began.

In 1859, Charles Darwin introduced the Theory of Evolution in his book, *The Origin of Species*, and for the first one hundred years of the theory's history, the popular belief was that the universe was eternal. The theory rested upon the primary logical foundation that given enough time, anything could happen—with emphasis on the word "anything"—including inert matter randomly organizing itself into complex eco- and biological systems that are far beyond our ability to intentionally reproduce today. That was until Mr. Hubble came along and knocked a hole in that foundation.

Science now says there is a universal time limit. Estimates are that the universe is approximately 13.7 billion years old. Current geophysical estimates are that the earth is approximately 4.5 billion years old. Given the fact that the universe is very complex and ordered—with one of the most ordered and complex things in existence being you and me—probability calculations show that the odds of us occurring by chance in that amount of time are simply zero.

In later chapters, we will see how current cosmological theorists have attempted to compensate for this mathematical flaw in the foundation by speculating that our entire universe is but one of innumerable universes—like a bubble in a bathtub full of universes with one entire universe in each bubble. Other theories postulate the existence of numerous other dimensions or tiny strings, all of which are far beyond the ability of modern science to actually detect.

However, as we will also see, the time and space inside of which these new theories are constructed are themselves anything but a reliable framework for the fertile imaginations of today's theoretical physicists.

CHAPTER 18

THE ULTIMATE QUESTION

The mystery of the beginning of all things is insoluble by us.
—Charles Darwin

THE THEORY OF Evolution does not address life's ultimate question: *Where did the information that makes us who we are come from?* Science deals with physics—physical evidence. By definition, it does not attempt to answer the question of where the instructions came from that make a tree shed its leaves or make us grow and function as we do. As far as where these instructions came from and how they got into inert matter and started making it move—become alive—no one has a clue.

When you want a computer to work, there are three basic requirements. First, you need the hardware—the hard drive, keyboard, monitor, and so on. Then you need energy: a power source. The final requirement is software or instructions for the energy to carry to the hardware so that it will actually do something. Without any instructions—any information—all you have is a lump of plastic and metal.

Science studies the power source—energy—and the hardware—elements—not where the information came from that is carried by that energy instructing the elements to behave as they do.

The energy that moves information through the hardware driving every single thing that happens in the universe is light. In the chapters to come, we will see how Albert Einstein began to uncover the incredible ramifications of this simple truth. However, science is silent as to where the information came from that causes the elements to grow, walk, and talk. It simply says that once the elements accumulated on this planet, life just appeared. Somehow, non-living matter began to move and organize itself into plants, animals, and the DNA that created you.[35]

The Bible is not silent as to where the information that created everything in existence came from. In fact, it says that before anything existed in the physical universe, there was first information.

> In the beginning was the Word [information], and the Word was with God, and the Word was God. He was in the beginning with God. All things came into being through Him, and apart from Him nothing came into being that has come into being. In Him was life, and the life was the Light of men.
> —John 1:1–4

The *Amplified Version* says "before all time was the Word" (John 1:1). So the Bible says that before time existed, life existed in information—"the Word"—and it was that information that constructed everything. The Theory of Evolution takes as a given that, despite the natural laws of decay, dirt and water randomly organized themselves into information far beyond our own ability to intentionally duplicate or even understand to this day.[36] Though on the surface there seems to be an inexplicably stubborn insistence within the scientific community to hang on to this idea, the number of scientists who are realizing this is impossible continues to grow.[37]

Some have paid a heavy price for not adopting the prevailing politically correct evolutionary view. But why? Whether one believes that God, evolution, or aliens explain the source of life, no one disagrees that before this planet, the life on it, or anything of any kind existed, there was no matter, no space, no time, until *bang*, there was. Even if one does believe that the elements on earth spontaneously organized themselves into people, current scientific theory still states that these elements and the stars they came from had a supernatural beginning.

Both the Bible and contemporary scientific theory state that, ultimately, what we are is beyond the ability of natural science to understand. The arguments in our courtrooms, on our school boards, and even within our families seem to be over something about which we all really do agree. No one denies that, when it is all said and done, the seed of our existence is, in fact, supernatural.

CHAPTER 19

THE THREE FUNDAMENTAL FORCES

> Magnetism is one of the fundamental forces in the universe,
> the others being gravity, duct tape, whining, remote controls,
> and the force that pulls dogs to the legs of strangers.
> —Dave Barry

DAVE BARRY HAS a unique way of looking at things. So did Albert Einstein. During the latter years of his life, our friend Al sought to understand the basic forces that are responsible for everything that occurs in the universe. Along with quantum mechanics, what his theory of relativity did was show physicists that there are three fundamental forces or energies that cause everything that happens in the cosmos. No matter what it is—black holes absorbing light in distant galaxies,[38] grass growing, your heart beating, whatever—all things that happen in the entire universe are caused by one of these three fundamental forces.

The first is *gravity*. In his general theory of relativity, our friend Al demonstrated that gravity was not a force that objects exerted on each other—as Newton had explained—but the result of mass bending space. You can picture this by imagining a bowling ball resting in the middle of a trampoline. If you place a tennis ball nearby, it will roll toward the bowling ball. This will not occur because either ball is exerting force on the other, but because the space around the bowling ball is curved, causing the tennis ball to roll toward it. Now replace the bowling and tennis balls with the sun and earth. The earth rolls around the curved space caused by the sun. However, because there is no friction in space, the earth never slows down or falls into the sun. It just keeps on rolling. This explains why objects of different weight fall at the same rate and why something without any weight—such as light—will follow the same path through space as something that has weight.

Most people thought our friend Al's theory was laughable until 1919 when light was observed following a curved path through space around the sun. It didn't take long, however, before an apparent hole was found in relativity's theoretical foundation. Experiments with atoms began to show that the law of gravity falls apart at the quantum level. The things of which atoms are made—quarks, electrons—do not obey the rules of gravity the way that you, I, and larger objects must.

Making sense of the relationship between the behavior of things in the quantum world and the rest of the universe is a fundamental mystery in physics today. If such a thing can be found, whoever does uncover it will, like our friend Al, become world-famous in an instant.

The second of the three fundamental forces that cause everything that happens in the cosmos is the *strong nuclear force*. In the world in which we live, positively charged objects repel positively charged objects. However, this is not the case inside the nucleus of an atom. There—and only there—positively charged protons will bind to each other. This is accomplished by what scientists refer to as the strong nuclear force. There is no natural explanation for why this extraordinary exception to a primary law of physics should occur only inside atomic nuclei except that if it were not so, we would all instantly vaporize.

The third fundamental energy or force that is responsible for everything else that goes on in the entire universe is the *electromagnetic force*. "Electro" is another term for "electric" or "light" energy. This is intriguing because—as we will see later—when the Bible describes the physical nature of God, it describes Him as "Light."

In the early nineteenth century, a gentleman by the name of Michael Faraday observed that magnetic and electric charges produce fields in the area surrounding them. By the middle of the century, a Scottish scientist by the name of James Clerk Maxwell noticed that the way these fields affect things in the area nearby is similar to the way water waves affect things. Imagine dropping a rock in a pond and watching twigs and leaves floating nearby bobbing up and down. Maxwell saw the same kind of disturbance around electric and magnetic fields and thus concluded that these forces travel in waves.

He was right.

Digging further, Maxwell also found something that, in retrospect, might have been the most significant discovery of the nineteenth century. Moving magnetic fields produce an electric current, while moving electric fields produce magnetism. What this meant was that electricity and magnetism were actually different aspects of the same force. So, being the ever-practical thinker he was, Maxwell called it "electromagnetism." He even measured the speed at which these electromagnetic—light—waves travel and found that they reach 186,282 miles per second. This led to the discovery that these light—or electromagnetic—waves come in an enormous variety shapes and sizes called "frequencies."[39] TV signals, microwaves, X-rays, and the red light you—almost—ran last week are all light—electromagnetic—energy waves whose unique abilities and characteristics—color, ability to cook things, carry a TV signal, see through to your bones—are determined by their different sizes and frequencies. The entire spectrum of light energy waves is called "electromagnetic radiation."

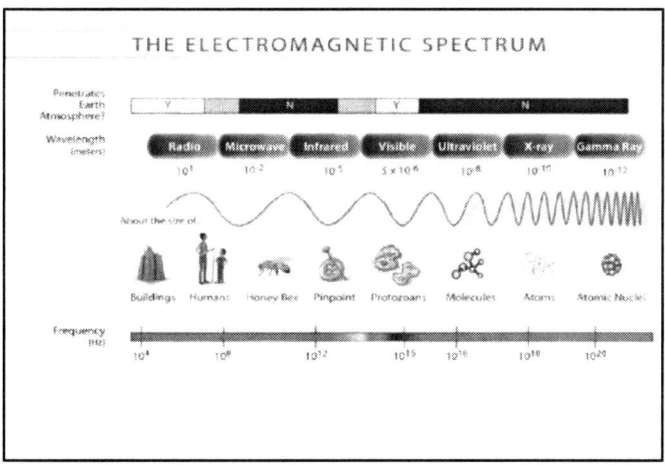

Most of us possess two instruments above our noses that are able to detect about 1/1000th of all electromagnetic energy waves. Visible light is but a small portion of all light—or electromagnetic radiation. The rest of the spectrum's radiation—the waves carrying cell phone calls, radio programs, TV shows, and so on—pass right through us undetected. In truth, there is much more to light than "meets the eye."[40]

CHAPTER 20

LIGHT CARRIERS

> The only thing that interferes with my
> learning is my education.
> —Albert Einstein

ISAAC NEWTON, QUITE logically, concluded that light consists of particles because it bounces off objects the same way other things, such as balls, do. Unhappy with the status quo, as the nineteenth century began, a man by the name of Thomas Young conducted an experiment that showed light had the characteristics of a wave. As we saw in the previous chapter, by the mid 1800s James Clerk Maxwell had also shown that light traveled in waves. Newton's ideas, however, were not easily tossed aside. He had been right about so many things that nobody was willing to instantly ignore his opinions on any matter.

The last half of the nineteenth century saw a lot of argument and experimentation concerning the nature of light. However, by the time the twentieth century turned, Maxwell's theories had been confirmed, and it became well established that light was a wave and that the different lengths and frequencies of the light waves do, in fact, make up the electromagnetic spectrum.

Then, just when all seemed ducky, our friend Al messed up everything and put physics into a quandary from which it has yet to recover. It seems that he noticed electrons being displaced by light on a photoelectric plate.[41] The only possible way this could occur would be if light consisted of particles as Newton had first surmised.[42] But wait!

The painstaking work and research of the entire nineteenth century had finally established that light was a wave. Therefore, what Einstein was proposing was clearly impossible. However, in the spirit of Galileo and Groucho Marx, our friend Al refused to let the facts—as the world plainly knew them—interfere with what was in front of his own two eyes. Suddenly, science had a dilemma. On the one hand, men such as Young and Maxwell had produced clear and unmistakable evidence that light was a wave. On the other hand, men such as Max Planck and our friend Al had produced compelling evidence that light was a particle. At this point, the scientific community—once again proving they weren't getting the big bucks for nothing—came to the forefront of intellectual inquiry and yelled, "*Okay, which is it?!*"

To find out, researchers—once again—went back to "square one." But when they did, something very unexpected happened. In fact, it was so unusual that, in all actuality, every notion of common sense reality was turned on its head.[43] And to this day, that is exactly where it remains.

Unwilling to deny what was in front of him, our friend Al proposed that light consisted of tiny particles—later called photons—and that they traveled in distinct packets—or quanta as Max Planck had earlier described these packets—of energy. It took a couple of decades, but other researchers gradually began to realize that, despite all of the evidence that light was a wave, what Einstein had uncovered could not be denied. Thus, the science of quantum mechanics was born. Quantum mechanics describes what goes on in the world of atoms.

In the previous chapter, we saw that light—electromagnetic radiation—is the means by which energy travels and that it is this transference of energy that is behind every single thing that occurs in the universe unless it is caused by gravity or is holding things together inside the nucleus of an atom. A burp, photosynthesis, rain, sunshine, a moving car, a thought—you name it—electromagnetic radiation—light energy—is the means by which it happens.

Peering into this quantum world of atoms, scientists saw that this energy transfer is accomplished by the exchange of photons—light particles—between atoms. So unless something is caused by gravity or energy transfers that occur only inside the nucleus of an atom, everything that happens in the universe is caused by the energy that results from atoms exchanging photons.[44] Simply stated, light—photons—is nature's way of transferring energy through space. The universe runs on light.

Reality Check

> As previously mentioned, this is one of the most important discoveries in history. However, it is possible that before this dawned on Mr. Einstein and his colleagues, there were a few others who understood this concept. It is possible that there were those who took the Bible's declaration that "God is light" (1 John 1:5) and that "He is the sole expression of the glory of God [the Light-being, the out-raying, or radiance of the divine . . . upholding and maintaining and guiding and propelling the universe]" (Heb. 1:3 AMP) at face value. Of course, this is a completely uneducated and un-theological thing to do—unless we believe, like our friend Al, that sometimes the only thing that interferes with our learning is our education.

As the twentieth century progressed, physicists began to better understand how the three basic forces—gravity, electromagnetism, and strong nuclear force—were at work shortly after the time of creation/Big Bang—whichever you prefer—and how all the energies and forces we know of are a form of one of these basic three. However, the force of gravity created a problem. As we saw in the previous chapter, gravity is nowhere to be found in the sub-atomic world of quantum physics. What occurs there is so incredibly strange that it baffled Einstein, Niels Bohr, and has continued to completely puzzle physicists ever since. How can atoms defy the law of gravity on a small scale yet obey it when they are clumped together in the form of you, me, and all the other junk that surrounds us?

It seems obvious that the three forces were created during the initial creation/Big Bang that brought everything into existence. So it would make sense that, for at least a millisecond or two after the Big Bang, they all existed as one unified force and therefore are related in some way. Physicists believe that from the incredibly intense energies that existed in the first few moments of the universe, a single fundamental force will be found that acts as these three separate forces at lower energies.[45] This idea is commonly referred to as the "Grand Unification Theory" or "Theory of Everything."

Despite the fact that scientists have been searching for that connection for almost a century now, no one has been able to find it. As it has been for the past one hundred years, the Theory of Everything remains the Holy Grail of physics today, for understanding it would mean understanding how everything in existence is related, and there is no telling what wonders such knowledge would unlock.

CHAPTER 21

FALLING APART

Things that seem to be solid are not.
—The Byrds[46]

IT WAS ONE oppressively hot day in Washington, D.C., when curiosity about the quantum world hit me over the head and wouldn't leave me alone. Like too many other days in my life, it started with me thinking I already knew just about all there was to know on the subject. I had formed this neat little picture of electromagnetism, gravity, and the strong nuclear force, and thought I understood it all. There I was walking around saying to myself, *I'm crusin' now, man. I have got it toooogether. I mean, I actually know what I'm talking about. I've got to try this more often.*

That lasted two weeks.

I was trying to keep up with my kids that day on the Washington Mall and was rescued by a large, free to enter, air-conditioned building. Sometime later, I found out it was the Smithsonian National Air and Space Museum.[47] Sitting down in a cool quiet place, I found myself in front of a short film about atoms. Before I could fall asleep, I was informed that there was something about their nature of which I was unaware. An atom is pictured like a little solar system with the nucleus—the collection of protons and neutrons—in the center and the electrons in orbit around it. And they are really small. There are as many atoms in an apple as it would take apples to make up the entire earth. There are billions in a grain of sand.

Suddenly, I was told something about these incredibly small things that caught me quite unaware. The nucleus of a single atom occupies only about one-ten thousandth of an atom's actual size.

"*Clang.*" That was the sound of my brain sliding off the road it was *crusin'* into a ditch. Then suddenly, the quiet in the theater was shattered as the words, "*Wait a second,*" blurted out of someone's mouth. At virtually the same moment, my daughter kicked me in the leg. Undaunted, the same voice could be heard loudly asking the narrator of the film, as if he could hear, "*But what else is there in an atom? Electrons? How big are they?*" As I rubbed the second sore spot on my leg, the narrator informed us that electrons are thousands of times smaller than a nucleus.[48]

Okay, let's take this slowly, I thought. *We were told that inside one grain of sand there are literally billions of atoms. Each one consists of electrons orbiting a nucleus. The nucleus makes up about one-ten thousandth of the size of an average atom and the electrons are thousands of times smaller than the nucleus. So what does that leave you with?*

Reality Check

> That's right, it leaves you with next to nothing, and don't act like you haven't been there before.

Moving this to a scale we are familiar with, what this means is that if the nucleus of an atom were the size of a marble, the electrons would be the size of a grain of sand in orbit around the nucleus … two miles away.

To sum up then, 99.999 percent of every atom is empty space. Yes, that means that 99.999 percent of every thing that exists is empty space. No, I'm not leaving you out. I know you always suspected this about some people, but the fact is we're all 99.999 percent empty space.[49]

Just in case you were wondering about the actual .001 percent of the atom (nucleus) that is (supposedly) not empty space, let's take a look at that for a moment. We see that the protons and neutrons that make up the nucleus are made of even smaller particles called quarks. Each proton and neutron has three quarks. But a closer look at quarks reveals that all three make up only about 2 percent of the mass of a proton or neutron. Not surprisingly, the mass of an electron is thousands of times smaller than a proton.

Are you saying that when you get right down to it, nothing at all is actually physically solid?!

Reality Check

Yep.

There I was. My entire ego and worldview shattered inside of thirty minutes one hot afternoon. Cautiously stepping on the escalator, I made my way out of the museum in the wobbly optimistic belief that whatever it was that stopped me from being sucked into the air conditioning on the way in was probably going to keep me in one piece on my way out. Outside, I wondered (as one might) what exactly it was that was preventing me from falling to pieces and leaking through the cracks in the sidewalk. Maintaining a tenuous hold on the impression that parts of me were not drifting away on the afternoon breeze, I deftly concluded there had to be an answer.

Brilliant!

Yes, but what is it?

Oh, you want details. That's different.

The details took a bit more time. A few years, a decade, maybe two—but who's counting? This book found its genesis on that afternoon.

So what is it that keeps everything together? If matter is not actually solid, what exactly is it that we perceive? Just what is this reality we inhabit?

It is the electromagnetic energy described chapter nineteen. It truly is responsible for everything that happens in the universe unless it is caused by gravity or the strong nuclear force.

When you slap your hand on the table, the protons and neutrons do not actually touch each other. It is the electromagnetic—light—force fields that comprise your hand and the table that bang against each other.

The outer shell of every atom contains negatively charged electrons. Like (the same kind of) electric charges repel each other. When your hand gets close to the table, the negative electric (light) charges that hold your hand together repel the negative light charges that comprise the table.

In other words, the reality we inhabit is light energy.

CHAPTER 22

WHO KNEW?

> The most important scientific revolutions all include, as their only common feature, the dethronement of human arrogance from one pedestal after another of previous convictions about our centrality in the cosmos.
> —Stephen J. Gould

AFTER JESUS LEFT this earth, His followers circulated His story by word of mouth. If you have ever had someone tell you a story and then heard that same story from someone else just a few minutes later, you are familiar with how much it can change in just one rendition, let alone after it has been passed around a few more times.[50] As the first century drew to a close, the church was growing and, inevitably, some misinformation about Jesus was circulating.

To correct some of these inaccuracies, the apostle John wrote a letter that is now preserved in the Bible as the book of 1 John. At the time, he is believed to have been the only one of Jesus' twelve disciples who was still living. John intended for his letter to be circulated among the churches in order to set certain things straight, and to do that, he needed to start at square one. His readers had to understand exactly *Who* Jesus was. With only word of mouth to go on, in many quarters that understanding had not by any means been clearly established, and it was crucial. If people did not clearly understand Jesus' true nature, then all that came after would be incorrectly interpreted and eventually fall apart due to a faulty foundation.

As we read the opening paragraph of the letter, it is apparent that John wanted his readers to plainly recognize Jesus' identity in no uncertain terms. He began like this:

> From the very first day, we were there, taking it all in—we heard it with our own ears, saw it with our own eyes, verified it with our own hands. The Word of Life appeared right before our eyes; we saw it happen! And now we're telling you in most sober prose that what we witnessed was, incredibly, this: The infinite Life of God himself took shape before us. We saw it, we heard it, and now we're telling you so you can experience it along with us, this experience of communion with the Father and his Son, Jesus Christ.... We want you to enjoy this, too.... This, in essence, is the message we heard from Christ and are passing on to you: *God is light, pure light*; there's not a trace of darkness in him.
> —1 John 1:1-5 MSG, emphasis added

First John 1:5 in the *King James Version* says, "This *then* is the message," meaning that as a result of all that John had heard, seen, and touched, the message, at its core, is that *God is Light*.

Before the theory of relativity and quantum mechanics, in the everyday world the term "light" simply referred to the thing that gave a person the ability to see. However, these two breakthroughs completely changed the meaning of the term. As we saw in chapter nineteen, today the definition includes all the wavelengths of the entire electromagnetic spectrum. In the previous chapter, we saw that it is *light* energy that holds us and everything else in the universe together. In chapter twenty, we saw that "photons," are *light* particles that carry messages from one atom to the next. Science tells us that these messages—instructions—set in motion every single thing in the universe unless the motion is caused by gravity or occurs only inside the nucleus of an atom.

For the past two thousand years, most scholars and theologians have interpreted this description of God as light as a metaphor or symbolic description of God's nature. However, if, just for the sake of argument, what John said in 1 John 1:5 was not meant to serve as an illustration but is to be taken literally, it would not be the first time a Bible passage was considered symbolic until a scientific discovery revealed otherwise.

Let's look at some other passages that for centuries were seen as symbolic representations of the spiritual realm—not a physical reality in the natural universe—until science proved otherwise.

Certainly, Job's statement that God "stretches out the north over empty space; He hangs the earth on nothing" (Job 26:7 NKJV) was metaphorical rhetoric to most readers until Copernicus and Galileo revealed that the old man actually knew what he was talking about. Old statues or pictures of an angel or some other sort of supernatural being holding up the earth still exist today.

Sometime between 765 and 750 B.C., a farmer by the name of Amos wrote, "It is the LORD who . . . draws up water from the oceans and pours it down as rain on the land" (Amos 5:8 NLT). Apparently, to be sure the point was not missed, four chapters later Amos again announced, "He who calls for the waters of the sea and pours them out on the face of the earth, the LORD is His name" (Amos 9:6). The book of Job is one of the oldest (if not the oldest) biblical texts, recording events that took place some four thousand years ago. Job 36:27-29 says, "He pulls water up out of the sea, distills it, and fills up his rain-cloud cisterns. Then the skies open up and pour out soaking showers on everyone. Does anyone have the slightest idea how this happens?" (MSG). The answer at the time was absolutely nobody knew how it happens. Two thousand years later (about five hundred years after Amos), Aristotle wrote that rainfall might be able to produce a river, but since there was so much water in lakes and rivers, he believed that the water supplying them must come from vast underground canals connected to the sea.

It was not until approximately A.D. 1500 (some 3,500 years after the author of Job marveled at the process and some 2,250 years after Amos left the scene) that the concept of evaporation came to someone else's mind.

That person was Leonardo da Vinci. Between painting famous works of art such as the *Mona Lisa* and *The Last Supper,* designing flying machines, messing with geometry, building churches and forts, and producing the first human anatomical studies, Leonardo, in his spare time, wrote about water evaporating from the surface and returning to the earth. However, like Aristotle, because of the vast amount of water in the rivers and lakes, Leonardo still held on to the notion that they had to be supplied from the sea via underground canals. In fact, as late as the mid 1600s, great maps were drawn of vast underground passages that stretched from the sea to the mountains to supply rivers.

For more than twenty-five hundred years, the words written by Amos, and for more than thirty-eight hundred years the words written in Job, were considered to be figurative references to the nature of God. Then around 1800, an Englishman named John Dalton became the first non-biblical writer to describe the hydrological cycle in detail. He suggested that water evaporates from the sea, forms clouds that are carried by the wind, and then pours out onto dry land through a process that, as stated in Job, *distills* (extracts or leaves behind) the salt in the ocean.[51]

As we saw previously, Isaac Newton unified what had always appeared to be totally unrelated events (planetary motion, tides, a falling glass) by showing that they were all caused by the same thing: gravity. We will examine later that this is what our friend Al did with space and time. He showed how they were aspects of the same thing and could not be separated.[52] When most physicists try to describe this today, they use terms such as the "fabric of space." They, however, are not the first. About twenty-seven hundred years ago a man named Isaiah asked: "Do you not know? Have you not heard? Has it not been told you from the beginning? … It is God Who sits above the circle (the horizon) of the earth … it is He Who stretches out the heavens like [gauze] curtains and spreads them out like a tent to dwell in" (Is. 40:21-22 AMP).

Likewise, the psalmist reported: "[You are the One] Who covers Yourself with light as with a garment, Who stretches out the heavens like a curtain or a tent" (Ps. 104:2 AMP).

Who would have thought that these statements were anything more than dreamlike hyperbole, until a few years ago when Einstein, Stephen Hawking, and others told us that the term "fabric of space" really is the best description of the true nature of our universe?

In chapter sixteen, we noted how the author of Hebrews 11:3 declared that time is not eternal but created, and how Paul asserted in Romans 4:17 that God creates new things out of nothing. As we have seen, before the advent of the Big Bang Theory, these statements were considered figurative symbolism and not written in reference to actual reality.

Regardless of what one believes about the subject matter, John's writings expose an intelligent, introspective man. Metaphorical imagery is not his subject. It is the very opposite. John states that his purpose is to convey the reality of what he personally heard, saw, and touched. His subject is the fact that the infinite Life of God Himself took shape right before his eyes. It is what he saw with his physical eyes, heard with his physical ears, and touched with his own hands. In other words, John's subject is physics.

But believe it or not, this is really just the beginning of the story. The heart of the matter still lies ahead.

CHAPTER 23

STAR STUFF

> I didn't fail the test, I just found one
> hundred ways to do it wrong.
>
> —Ben Franklin

AS WE SAW in chapter 21, when you press your hand down against the top of a table, there is no actual physical matter contacting one thing or the other. What is happening is that the electromagnetic (light) force holding you in one piece is being repelled by the electromagnetic force holding the table together. This is the true nature of reality. It is a fundamental aspect of relativity and quantum mechanics.

To look at it another way, consider the following story. Etched in my memory, for some reason, is the day back in seventh grade when I first caught a glimpse of a large chart with something on it called the "Periodic Table." It came to my attention late in the afternoon after I tripped over the thing in front of twenty-four hysterical seventh graders who had been waiting all day for some comic relief. Just to be sure I had my facts straight, I recently checked with an old friend who assured me that, yes, it was I who fell into the chart.[53] He then told me that for some strange reason, ever since that day he has actually remembered that the chart showed that atoms are made up of protons, neutrons, and electrons, and that each of the ninety naturally occurring elements in the universe have a different number of these. These ninety atoms make up or combine to make up everything. Not just most things, but every single thing in existence.

Back in chapter seventeen, we saw that most of these elements were forged inside of dying stars. As Carl Sagan so famously said in the TV series *Cosmos*, we are all made of "star stuff." Or to put it another way, we were made inside a light machine (a star). However, not only were we made by light, but when you get right down to it, that's actually what we are. This is because everything in the universe is made up of atoms, which themselves are made of protons, neutrons, and electrons. A proton is a positive electric charge (or "light energy"). An electron is a negative electric charge. A neutron, of course, is neutral.

When you look at the periodic table, the first thing you notice is that each atom on the chart has the same number of protons and electrons. This is significant because of the way in which electrical charges react to each other. Back in 1750 when Ben Franklin flew a kite, he became one of the first (if not the first) to observe that like (or the same kind of) electrical charges *repel* each other while opposite electrical charges *attract*.[54] He then showed up late at the Continental Congress a little fried around the edges. All of this went to show that in addition to the fact that all people are endowed by their Creator with certain inalienable rights, electricity—light energy—is the reason those so endowed stay in one piece. Each element remains stable due to the fact that it has an equal number of positively charged protons and (opposite) negatively charged electrons.[55]

CHAPTER 24

PRECISELY BALANCED

> A child of five would understand this.
> Someone find a five-year-old.
> —Groucho Marx

WE NOW KNOW that atoms are very small and that, despite their size, they are stable due to the equal number of positively charged protons and negatively charged electrons in each one. What's amazing, however, is just *how* precisely balanced they truly are. Scientists have measured the charge of a single electron and a single proton and have found that they are *exactly* equal and opposite. When I say "exactly," never has the term been more meticulously or strictly applied. If the charges of the unimaginably small electron or proton differed by as much as one part in one hundred billion, every single person, animal, plant, rock, or molecule would instantly either fall into nothing or explode like a nuclear bomb.

Back in seventh grade, the day after I tripped over the periodic table in front of twenty-four still-giggling seventh graders, the teacher told us that the way atoms build everything is by combining in different ways to form molecules. For example, when two highly combustible things like two hydrogen atoms and one oxygen atom get together, they form something very refreshing and useful[56] called water.

At this point, you may well have surmised that the only thing I was ever able to comprehend about chemistry was that it is pretty easy to stink up the bottom floor of any high school or collage in pursuit of one of the primary reasons for attending any institution of higher learning: humor. However, just a brief look at the number of ways in which these elements can be used and combined is mind-boggling. It appears, in fact, to be never-ending. The same elements have been here since the world began, and people are still heating, freezing, frying, and combining them into a never-ending array of stuff. One trip to Wal-Mart is proof of that.

In a way, the elements in the periodic table are pieces of information much like the twenty-six letters in the alphabet. People have used these letters to create every book ever written. Each one of these works was made with the same building blocks, but each one is nonetheless unique. So in addition to the fact that electromagnetic energy—or light force—holds everything together, we see that what makes each thing—glass, grass, your rear end, and so on—unique is its different electromagnetic—light—energy configuration.

CHAPTER 25

THE QUEST TO UNDERSTAND LIGHT

> The most beautiful thing we can experience is the mysterious.
> It is the Source of all true art and science.
> —Albert Einstein

AT THE BEGINNING of the twentieth century when our friend Al first wrote down a small equation ($E=mc^2$), most people were not living any differently from their ancient ancestors. A few had nice houses with interesting forms of indoor plumbing; but then again, so did the ancient Egyptians. Most of the earth's inhabitants used legs for transportation; fire for light and heat; blood, sweat, and tears for work; and had a real adventure on their hands if they needed to take advantage of a bathroom on a cold winter's night.

This is why the theories of relativity and quantum mechanics were such breakthroughs. Actually, they were more than breakthroughs—they were scientific revolutions. In relation to all of human history, the understanding of relativity and quantum theory almost instantly took people off donkeys and put them on the moon. In one generation the human race went from megaphones to cell phones, from postcards to the Internet, and beyond. Relativity and quantum mechanics have changed everything about life on this planet we call home.

These discoveries came about as a result of the quest to understand *light*. When Job began questioning God's integrity nearly four thousand years ago, God silenced his arrogance by asking him, among other things, if he knew where light came from (see Job 38:19). Job didn't have a clue, and despite all of the progress we have made in this quest, we still know very little. Roger S. Jones, Professor Emeritus at the University of Minnesota, put it this way:

> There are mysteries in the nature of light that have eluded all the explanations of science. How does light originate and travel through space? ... What is light—substance, vibration, pure energy? Is color fundamental to light or merely a matter of perception? Science has found some partial answers to these persistent and challenging questions, but at the deepest level, they remain unanswered (and perhaps are unanswerable). Yet, the quest to understand light has given rise to the two great revolutions of twentieth-century physics—relativity and quantum theory.[57]

Although there remain many "mysteries," a great amount of truth has been uncovered by the theories of relativity and quantum mechanics. The proof is everywhere. If quantum theory were false, people would be laughing at you sitting there staring at the TV with some crazy notion that a picture was actually going to fly though the air and appear in your house. It would make more sense to see someone holding a cup up to his ear instead of a small lump of plastic, expecting to hear someone on the other side of town, let alone the planet. Without a doubt, the world has been transformed by this "quest to understand light."

CHAPTER 26

THE ESSENCE OF LIFE

The energy of the mind is the essence of life.
—Aristotle

WE CAN'T SEE anything without light.

Reality Check

Duh!

Okay, let me put it another way: we can't see anything *but* light. Light is what enters the lens of our eye and hits the retina, which then sends an electrical signal down the optic nerve to the visual cortex in our brains. Although we say, "I see the wall" or, "I see the table," what we are actually seeing (or sensing) are the photons and/or, light waves reflecting off of those things into our eyes and decoding that information in our brains. Just as it is with our four other senses, the portion of our bodies designed to detect the stimulus does so and then sends the information somewhere between our ears for processing.

One thing that is interesting about this process is that it takes time. It takes time for light to make the journey from the object that emits or reflects it to our eyes and then on to our brains. At 186,282 miles per second, it takes light one billionth of a second to go one foot. That means, technically speaking, that you are always looking into the past. However, if someone pulls out in front of us in traffic, the amount of time it takes the light to reflect off of the oncoming car into our eyes is so small that it will not affect the outcome. You can compare it to standing on a scale and having a speck of dust land on you. But still, the process does take *some* time.

Reality Check

> This was a fact that our friend Al could not ignore. He saw what was not apparent to the naked eye and uncovered the ramifications of this truth, and it was his singular vision that became the foundation of the theory of relativity. As we continue to unravel the nature of the universe and our place in it, we find that understanding the fact that light cannot travel through space without taking up some time to make the journey is absolutely fundamental to our understanding of the reality we inhabit.

When our bodies detect a stimulus and send it to our brains, it does so through electrical impulses: light. In this sense, you could say that each of us is a "light machine." Our central nervous system is a myriad of electric circuits, and like any electrical system, it consists of something for the signal to flow through. Instead of wires, however, we have nerve fibers and a lot of them. In fact, if we were to stretch out all the nerves in our bodies in a straight line, they would go around the world and on again for another two thousand to three thousand miles. We are a huge and unimaginably complex electrical—light—system. Every animal is. Every time an animal moves, it produces a weak electrical charge. We, as human beings, are not aware of this, but sharks are. Every shark has hundreds of thousands of tiny electricity receptor organs in its skin that detect the electric (light) charges produced by other animals. The scientific term for each individual receptor is an "Ampullae of Lorenzini." Hammerhead sharks are the most light-sensitive animals on earth. They can sense electric fields produced by animals hiding under the sand.

All electrical systems have switches that either allow light to flow through the circuit or turn it off. The switches inside us are called neurons. Our bodies contain about two hundred billion of these switches. However, unlike most manmade circuits, which contain only one switch to turn them on or off, each single neuron in our body is capable of turning on or off about one thousand different circuits.

Imagine it this way. When you take a step, the neurons in your body turn on to establish a connection from your brain down to your foot. The information then travels back up your leg to your brain, which enables you to move and perceive that walking is, in fact, what you are doing. Many of your body's functions are controlled by this electrical system, whether you ever think about it or not. For instance, on top of your heart is something called the sinoatrial node, or pacemaker. When you were about the size of an apple seed in your mother's womb, it began sending electrical (light) impulses down your heart, and it has continued to beat about a hundred thousand times per day ever since. Numerous other body functions—such as breathing and blood pressure—are regulated automatically by light impulses from one nerve to the next.

When it comes to your brain, it is difficult to comprehend the amount of electrical activity that is taking place in there. Half—about one hundred billion[58]—of your body's neurons (switches) are located inside that three-pound lump of mush between your ears. Every thought you have ever had and every movement you have ever made is an electrical—light—event that originated in your brain, which, despite modern technology, remains far and away the most complex thing in the known universe. A pinch of human brain tissue smaller than the period at the end of this sentence contains about one hundred thousand neurons capable of making about a billion connections. That translates to about a quadrillion bites of data, which is enough to store about 50 years worth of high definition movies playing 24/7.

When you came to the end of the last page, you somehow enabled the neurons in your brain to send signals to each other to coalesce the idea that the page should be turned. Then you were able to send an electrical signal through just the right neurons (switches) to cause your hand and fingers to do the job. Each nerve generated approximately one-tenth of a volt of electrical (light) energy as it passed the signal through each switch from one nerve to the next.

This electrical system is important. In fact, it is our life. When death seems imminent, physicians typically monitor a person's heart activity with an electrocardiogram (EKG). This machine senses the electric—or light—energy that causes that person's heart to beat. If there is no light, there is no heartbeat. Physicians also monitor the electrical activity of a dying person's brain with an electroencephalogram (EEG). When either an EEG or an EKG "flatlines"—meaning that no more electrical activity can be detected—physicians often list the cause of death as brain or heart failure. However, in the minute or so before damage sets in due to lack of oxygen, the heart muscle is still in working order, and so is the brain and every other organ in the body. If the electrical (light) activity can be restarted within a short period of time (using CPR or a defibrillator), a person can continue to live a normal life for many years. If not, the person's organs can be harvested during this short period of time and function normally in another person's body for many years. The issue, then, is not heart or brain failure; it is light.

Reality Check

"There it was—the true Light coming into the world that illumines every person."
—John 1:9 AMP

"In Him was life, and the life was the Light of men."
—John 1:4

Of course, the eternal question is what happens to this electrical light energy when it leaves a body. One of the most basic laws of physics is called the "Conservation of Energy." It states that there is a fixed amount of energy in the entire universe that will never change. The amount of energy released at the beginning of everything is *exactly the same* as the amount of energy that exists right now. What changes is the form that it takes. This means that energy—or the ability to do things—can never be destroyed or created; it can only change places and forms. So according to one of the most fundamental laws of physics, the energy—light—that is you will leave your body someday, but it will never stop existing.

CHAPTER 27

NOTHING STAYS THE SAME

> Why shouldn't truth be stranger than fiction?
> Fiction, after all, has to make sense.[59]
> —Mark Twain

IF YOU KNEW a kid who couldn't resist smashing bottles on rocks or smashing anything with just about anything else, chances are that he or she grew up to be a physicist. Physicists are so into smashing things that they spend billions of dollars to build huge contraptions for that very purpose and then spend years waiting in line for the opportunity to smash those things into as many pieces as possible.

Reality Check

Let's stop here for a second and pretend you are back in school. Here's a question: In the years these physicists have spent smashing things, what do you think they have found? I'll bet you didn't raise your hand because you think you don't know. But you are wrong: *you did know.* I am sure it was the first thing that came into your mind. What they have found is that just like when you throw a bottle against a rock, the harder you smash something, the more

> pieces it flies into. It's just that they have been doing this for a long time now, and the pieces have become so small that they need to build expensive machines to see them. (See? You really do have a scientific mind.)

In the early days, these machines were called "atom smashers." Now, primarily because it is hard to get politicians to spend money on something called a "smasher," they are called "particle accelerators" or "Hadron Colliders." Hadrons are particles that make up atoms. Regardless of what they are called, the process is the same: these machines accelerate parts of atoms to a great speed and then—that's right—smash them.[60] This is done in huge underground circular tubes that are miles long. The latest is a 16.5-mile tunnel located one hundred feet under France and Switzerland called the Large Hadron Collider, which was constructed at a cost of more than ten billion dollars.

Let's take a look at what, so far, are the smallest particles atoms have been smashed into. Remember that the nucleus of an atom contains protons and neutrons. Physicists have smashed these two things into so many pieces over the past decades that the results have been termed a "particle zoo." Touring this zoo, we find that everything in existence is made up of the following:

1. Leptons: An electron is a type of lepton. As you know, an electron orbits the nucleus of an atom. There are five other leptons involved in radioactive decay that operate only inside the nucleus of atoms: electron neutrinos, taus, tau neutrinos, muons, and muon neutrinos.
2. Hadrons: Hadrons combine to make protons, and there are at least three hundred of them. However, hadrons have been found to be made up of something even smaller. They are called:
3. Quarks: There are six types of quarks. The first four are the "up quark," "top quark," "bottom quark," and "down quark." I suppose we would be left wondering what the difference between

"up" and "top" or "bottom" and "down" could possibly be if they hadn't named the other two the "charm quark" and the "strange quark." Each type of quark (up, top, bottom, down, charm, and strange) comes in three colors (red, green, and blue), for a total of eighteen different quarks.[61]

So there you have it. Everything in existence—you, me, this book, your dinner, your car, the sun, the moon, that bug—are all, when you get right down to it, made up of six leptons, eighteen quarks, or any combination thereof. That's it.

There is, however, one other group of elementary particles in the universe. They are called bosons. These particles are not involved in the composition of material things but are "messenger particles." Bosons are known to facilitate and communicate two of the three basic forces we discussed in chapter nineteen: electromagnetism and the strong nuclear force. The debate continues about exactly what facilitates the third force, gravity. Because every other known force is transmitted by messenger particles, it is assumed that gravity is transmitted that way as well. It is also assumed that, due to the fact that gravity is much weaker than the other forces, the imagined particle—graviton—is so small that it has yet to be detected.

As we discussed earlier, gravity, electromagnetism, and strong nuclear force are responsible for everything that happens in the material universe. Leptons and quarks are the universal hardware, while bosons carry instructions inside and between atoms telling the hardware what to do. There are twelve bosons and, like leptons, all but one work inside (and only inside) the nucleus of an atom. Three bosons, called W^1, W^2, and Z, carry radioactive decay messages to protons and neutrons. Eight of the others are called "gluons" and, you guessed it again, that is because they glue protons and neutrons together.

That leaves one other boson. This boson performs the same function as the other eleven, but it does not carry messages inside atoms. Instead, it carries messages outside of them, from one atom to another. What sends the message to the seawater, telling it to evaporate, rise up to the cooler air above, condense, and then fall back down on dry land? What carries the message to tell a tree to grow, the sun to shine, and your brain to think and wiggle a toe? What sends the message to every single atom in every single thing to change or do anything? Modern particle physics states that this messenger is a boson, or, more specifically, a particle called a "photon." This term is derived from the Greek word *phos,* meaning "light." In other words, each and every thing that happens in the universe is caused by light.

CHAPTER 28

THE HARMONY OF THE WORLD[62]

> As a scientist I believe that nature is a perfect structure, seen from the standpoint of reason and logical analysis.
> —Albert Einstein

WHEN ATOMS INTERACT, they gain and lose electrons. As this happens, the electrons jump from an orbit in one atom to an orbit in another. Sometimes they also change the position of their orbit in the same atom. Looking at all of this, scientists noticed something quite strange. It seems that electrons are extremely picky about exactly where they will orbit a nucleus of an atom. Although no one knows exactly why—except for the fact that nothing could exist if it were not so—electrons will only orbit a nucleus at certain specifically allowed distances. For instance, a hydrogen atom has just one proton—positively charged light force—and it has one electron—negatively charged light force—that orbits the proton. The charges are exactly equal and opposite and the electron circles the proton at one and only one specific orbital distance. A sulfur atom has sixteen protons and sixteen corresponding electrons. They occupy three different specific orbital distances or "shells," as science calls them.

Atoms vibrate and vibrations produce waves. We will find out later that all really small particles—photons, atoms, and the things they are made of—have a dual nature. That is, sometimes they are viewed as particles and sometimes they are measured as traveling in waves.

Reality Check

> If that doesn't make a lot of sense to you, don't worry about it. Remember, this particle-wave dual nature of matter was first discovered in the 1920s, and the reason very few have even heard about it after all this time is exactly that. It makes no sense . . . to anybody. As we have seen, Einstein, Bohr, Feynman, and the greatest minds of the past century could not make any logical sense out of it. That puts you right there in the same boat with them. It really is like just totally crazy. So now that we are 100 percent certain of that, we are free to:

Imagine an electron as a surfer who is looking at the waves of an atom. When a surfer is looking for a wave, she will only get on certain ones because they are the only ones that have the characteristics that will allow her to accomplish her goal. The same is true of an electron. An electron will only jump onto certain wavelengths (orbits) because they are the only ones that will allow the electron to give the atom the ability to become hydrogen, for example, where it has one choice, or sulfur, where it has three, as we saw in the above paragraph.

Not only do waves carry surfers, but they also carry sound and that also serves to illustrate the way electrons behave. The string on a guitar is tuned to a certain wavelength. The same is true of elements. In fact, it is because each element has a distinct vibration or frequency that astronomers are able to identify the composition of distant stars and galaxies. A plucked guitar string can only emit its fundamental *D* sound and its accompanying harmonics because it has a different length and thickness and therefore vibrates at a different rate than the *A* or *G* strings next to it.[63] The atoms in each element are tuned in the same way. An atom's electrons can only occupy an orbit or wavelength that is in harmony with the fundamental frequency of each element.

To complete the illustration, if an electron is in a low orbit at a *D major* frequency, it can move to a higher orbit in the same atom or an orbit in a different atom, but only if the wavelength of that orbit is in harmony with the *D major* wavelength. It cannot occupy an *E flat* or a *B flat* wavelength. If it could, music would not exist. Of course, that wouldn't matter much, because nothing else would exist either.

In chapter twenty, we talked about how atoms emit and absorb photons, thus propelling each and every change in the universe. Another way to look at this starts with the fact that matter absorbs light. If you try to shine a flashlight through a wall, you won't see much on the other side because the matter (brick, wood, plaster) will absorb the light. If you shine a light at a thin piece of paper, some of the light will get through, while some will be absorbed. Looking back at the electromagnetic spectrum chart in chapter nineteen, you will see that the higher the frequency of the light waves, the more energy the light will have to plow through things before it is absorbed. X-rays can see all the way into your bones.

Matter, of course, is made up of atoms, so it is the atoms in the wall or in the paper that absorb the light. More specifically, atoms absorb photons. If a photon contains the exact amount of energy an electron requires to move it to an allowed higher energy orbit (shell), the electron will absorb that photon and move to the higher orbit. If the photon contains the right amount of energy to move the electron two or three orbits higher, it will do that as well. Because a specific amount of energy vibrates at its own frequency, the photon's wavelength will be in harmony with the higher orbits. If a photon's energy is only sufficient to move an electron somewhere in between the orbits, the photon's wavelength will not be in harmony with any allowed orbits and the electrons will not absorb it. The photon will just pass by on its journey to the next block or next galaxy.

When you pluck a guitar string, it is the different tunings that cause you to hear a *D* sound as opposed to an *A* sound. The same is true of electrons. Electrons are tuned to occupy only certain energies or frequencies (harmonies), which cause you to see blue in one place and red in another. All the blue (*D*-string) frequencies resonate to let you see blue (hear *D*), and the red (*A*-string) frequencies do the same.

So not only is the fact that electrons are tuned to only certain energies or frequencies the reason we are here, but it is also the reason atoms construct the ninety naturally occurring elements and only those ninety elements. There are an infinite amount of other possibilities; however, because the universe adheres to this precise harmony, instead of the chaos that billions or an infinite amount of electron orbits and proton structures could create, ninety—and only ninety—elements make up a material universe that moves and changes in the orderly fashion that it does.

Because atoms and electrons vibrate in harmony, we are able to see blue and hear *D*. It is the reason freezing water forms ice and nothing else. It's the reason your eyes blinked. It is the reason you thought and then turned the page. It is the reason everything happens as it does.

Quantum theory, therefore, does not describe a world in which things are left to chance. It describes an extremely precise universal harmony.[64] The result is a stable and consistent material world. It is this precision, for example, that organizes, in exact detail, billions of bits of information into two cells that, when joined in a womb, produce a human being nine months later and nothing else. At the level of the atom, nature's most basic component, light—the photon—is the messenger particle that conducts a universal symphony.

CHAPTER 29

QUANTUM LEAP

> I think I can safely say that nobody understands
> quantum mechanics.
> —Richard Feynman

IT WAS NIELS Bohr who first proposed the idea of an atom in which electrons orbit around the nucleus at only certain allowed distances. This, along with his discovery that electrons changed orbits by absorbing and emitting photons, became the basis of today's quantum physics. For these reasons, Bohr and Albert Einstein are considered to have contributed more to modern science than any other individuals.

Bohr's understanding of the atom led researchers to another apparently simple question. Attempting to answer it, however, led physicist Richard Feynman to thoughtfully make the remark that serves as the opening quote for this chapter. This seemingly simple question is: "How do the electrons get from one orbit to another?" The answer is that no one knows.

Electrons appear to change orbits instantly. When I say this, it is not just to illustrate the point—I mean it literally. Electrons seem to disappear from one orbit and instantaneously appear in another without taking *any time whatsoever* to make the trip. And because they accomplish this in no time whatsoever, they also conveniently arrive at their destination without bothering to actually travel through any of the intervening space in between the two orbits.

As light is transferred between electrons, the atoms of which they are a part constantly change, combine, break apart, recombine, and so on to form elements, molecules, things, you, me, everything. As electrons change orbits, they somehow appear to exist outside of time and space. They vanish from one orbit and instantaneously appear in another by making what has come to be known as a "quantum leap." Again, this idea is not in conflict with the Bible's assertion that God is light and that He exists outside the constraints of space and time.

CHAPTER 30

THE GOD PARTICLE

> If we knew what it was we were doing,
> it wouldn't be called research, would it?
> —Albert Einstein

IN CHAPTER FOURTEEN, we discussed how, in order to explain the movements of the cosmos, scientists now theorize that 96 percent of the entire universe must consist of what is termed "dark matter" and "dark energy," even though, to date, no one has actually detected direct evidence of either one. Another fundamental mystery is how the other 4 percent of the universe—the matter we actually know about—came into existence. How did the light energy of the initial Big Bang become the helium and hydrogen that eventually fused to form stars? From our human perspective, logic would suggest that all of that radiation should have remained just that—diffuse light energy. How the light in the early universe took on the characteristic of mass that eventually turned into you and everything around you remains a mystery.

Back in the 1960s, an English physicist by the name of Peter Higgs suggested there might have existed another messenger-carrying particle—boson[65]—that, a few milliseconds after the Big Bang, instructed light to form or take on the characteristics of matter, and that this boson may continue to do the same thing today. Physicists have been looking for evidence of this "Higgs Boson" ever since. It has also been referred to as "The God Particle." The Large Hadron Collider mentioned in chapter twenty-seven was built as a direct result of this search.

Physicists use this huge collider to accelerate parts of an atom in one direction at 99.99 percent of the speed of light, accelerate more parts of an atom in the opposite direction at the same speed, and then smash the two together. By doing so, the idea is to recreate the incredible high energy and heat that existed shortly after the Big Bang to see if the resulting smash produces anything resembling a "Higgs Boson."[66] Who knows what they will find. One certain result will be the endless number of people signing up and waiting in line for years for a chance to smash something … anything.

CHAPTER 31

THE EXPERIMENT THAT WON'T GO AWAY

If stupidity got us into this mess, then why can't it get us out?
—Will Rogers

IN THE SEVENTEENTH century Isaac Newton understood that a wave is a surge of energy that requires something to carry it from one place to another. At the beach, you can't have waves unless you have water for them to move through. Sound is also a wave, and like the waves of the ocean, sound too must use something to get from its source to our ears. In our day-to-day lives, sound waves use air molecules to travel at about 769 MPH.

However, sound waves don't have to use just air to get around. In the ocean, sound moves closer to 3,300 MPH because the water molecules are much closer together. It goes even faster through solids. But without a mode of transportation (something to move through), there is only silence. No matter how hard you yell, without air, water, or something to carry the sound, the person standing next to you will not hear a thing.

Besides staring at waves a lot, Sir Isaac also correctly figured out that space was a vacuum (or close to it). Knowing these two things, he reasoned that light was made of tiny particles because (1) it bounces (reflects) off things the same way a tiny ball would, and (2) since space was a vacuum, there was no possibility that light could be a wave because space didn't contain anything that could transmit waves.

I am perfectly serious when I say that I am blown away by this man's impeccable reasoning. However, as has been the case with every great thinker in history (not to mention the rest of us), the truth continued its policy of having absolutely no regard for anybody's—even Sir Isaac's—flawless logic. Newton's picture began to unravel about two hundred years ago when a gentleman by the name of Thomas Young (mentioned in chapter twenty) performed a little experiment. Although the experiment was not all that complicated, the result was more than a bit interesting. In fact, what Mr. Young saw proved to be so fascinating that his experiment (or variations of it) is still being performed today in college physics classrooms all over the world. It is called the double-slit experiment.

What Young did was shine light on a screen with two very small slits (holes) in it. He placed another screen behind the first, which the light coming through the holes illuminated.

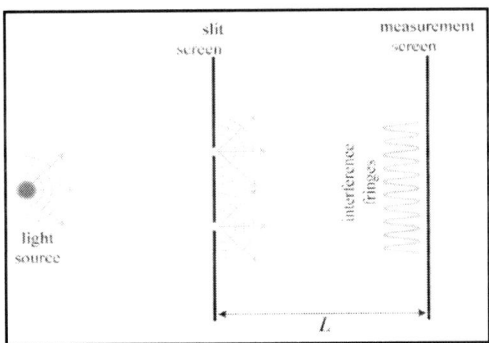

Young figured that if light consisted of tiny particles, as Newton had surmised, he would see a brighter area on the second screen where the light coming through both holes overlapped. But that is not what happened. What Young saw when he examined the second screen was light and dark areas.

Imagine dropping two stones into a pond at the same time and watching the waves move out in circles. Halfway between the two spots where the stones hit the water, you see the waves meet up and begin to *interfere* with each other. Where the crest of a wave from one stone meets the trough (low point) of a wave from the other, the two cancel each other out, and the water is flat. Where the crests of the waves from both stones meet, they combine to form a single larger wave.

The light and dark areas on the second screen showed the same pattern. Mr. Young saw that light behaved just like a wave. The dark areas indicated the places where the low and high point of the light waves coming through each hole had cancelled each other out. The bright areas showed the places where the crests of the light waves from each hole had combined to form a larger single wave. If light consisted of tiny particles, it would only produce a brighter area on the second screen—much like sand would pile up higher in the place where the grains coming through both holes would overlap.

When Young saw this pattern, he concluded that Newton had been wrong, and when he announced these findings, as you could imagine, he was not warmly received. His colleagues believed in particles, as had their forefathers. After all, they had based their careers on Newton's ideas, and it is not conducive to success in any field to go, "Oops, you know all that stuff I've been teaching for the past twenty years? Well, never mind." So they quickly went about the business of proving that Young was an idiot (though in scientific terms).

However, just as it has been throughout history, the truth continued its policy of having no regard for unanimous opinions or stellar careers. Despite countless meetings and votes to the contrary, the results of Mr. Young's experiment refused to change. In fact, to this day, there are still people who are very unhappy with what he found. As I said, this experiment continues to be performed today. In fact, in the two hundred-plus years since Mr. Young first fired up a few big lamps, the experiment has been executed so many times and with so many different variations that it is impossible to even begin to estimate the number of times it has been conducted.

One thing, however, has remained constant: despite two centuries of effort, hope, meetings, conventions, theories, and variations of the theme, whenever anybody shines a light at two small holes, wouldn't you know it, the results never change.

CHAPTER 32

THE LIGHT IS ALL

A light shines through us upon things, and makes us aware that we are nothing, but the light is all.
—Ralph Waldo Emerson

THE NEW TESTAMENT was originally written in the Greek language. In it, the Greek word most often used to describe "light" is *phos*. There are other words that translate as "light," but in this case, the word *phos* means actual light itself: the light that is emitted from the sun, a candle, or other source of illumination and enters our eyes. The word *phosphorus*, which is a chemical that ignites, emits light, and is used in matches, is derived from the Greek word *phos*.

In chapter nineteen, we saw that in the twentieth century, scientists uncovered the fact that the entire electromagnetic spectrum is, indeed, light energy. In chapters twenty and twenty-seven, we then saw how light—the exchange of photons—is the method by which every atom, and therefore everything in the universe, changes. In chapter twenty-two, we discussed how, if we read John's declaration that "God is light" (1 John 1:5) in the context in which it is used,[67] John is saying that God is actual light itself. To do this, he used the word *phos*.

Here are some other instances in the New Testament where the word *phos* is used to describe God.

1. "[God] dwells in unapproachable *light*, whom no man has seen or can see" (1 Tim. 6:16).[68]
2. "While he [Peter] was still speaking, behold, a shining cloud [composed of *light*] overshadowed them, and a voice from the cloud said, This is My Son, My Beloved, with Whom I am [and have always been] delighted. Listen to Him!" (Matt. 17:5 AMP).
3. "And the city [the new Jerusalem] has no need of the sun nor of the moon to give *light* to it, for the splendor and radiance (glory) of God illuminate it, and the Lamb is its lamp.... And there shall be no more night; they have no need of lamplight or sunlight, for the Lord God will illuminate them and be their *light* (Rev. 21:23; 22:5 AMP).
4. "The one who is the true *light*, who gives *light* to everyone, was coming into the world" (John 1:9 NLT).
5. The *light* that literally "illumines every person" (John 1:9 AMP).
6. "For God, who said, '*Light* shall shine out of darkness,' is the One who has shone in our hearts" (2 Cor. 4:6).
7. "I am the Light of the world" (John 8:12).

James, the brother of Jesus, used the word *phos* when he wrote, "Every good gift and every perfect (free, large, full) gift is from above; it comes down from the Father of all [that gives] light [*phos*], in [the shining of] Whom there can be no variation (James 1:17 AMP). Notice here that after James identifies God as the "Father (source) of all light," he then says that there can be no variation in its shining.

For the next nineteen hundred years, this text was thought to be a figurative reference to the moral character of God, and certainly that cannot be argued. However, as the twentieth century got underway, someone else decided to announce that there can be no variation in the shining of light. His name was Albert Einstein. In the chapters ahead, we will see that this was, in fact, the key that unlocked the secrets of relativity and quantum mechanics. Not only had James recorded that God is the originator (Father) of light, but he had also confirmed the characteristic of light that, once understood by physicists, launched the modern technological age that we live in. *There is no variation in its shining.*

CHAPTER 33

SEARCHING FOR AETHER

> When all possible answers have been eliminated, then the impossible is possible.
> —Aristotle

EARLIER, WE DISCUSSED Newton's hypothesis (a term used by scientists for the word "guess") that light consists of tiny particles. However, in chapter thirty-one, we saw that Thomas Young found that light travels in waves. James Clerk Maxwell later confirmed that finding. So it appeared that Sir Isaac had been wrong.

Maxwell's equations showed the speed of the waves of light to be constant at 678 million MPH (about 186,282 miles in one second). Our friend Al's equations showed that to be the universal speed limit. He said that it was not possible for anything to go faster than the speed at which light travels.

Also, we saw that waves are not physical things—they are simply the means by which energy flows from one place to another. When you see waves in the ocean, the physical water does not go zooming off with the wave, as evidenced by a boat or a bottle floating on it. As the wave of energy flows through the water, the bottle or ship bobs up and down in roughly the same spot.

So when scientists looked up at the light coming from the stars, they took Maxwell's measurement of light as a wave, combined it with the fact that waves require something to flow through, and concluded that Newton must have been wrong and that outer space must not be a total vacuum after all. *Something* had to be out there to enable the light waves to travel through it. This—being such a basic principle of physics—meant that even though no one had actually detected anything in space for light to move through, its existence was so obvious that they gave it a name: aether.[69]

In the late nineteenth century, two gentlemen named Michelson and Morley set out to find tangible evidence of this aether. It seemed an easy enough proposition, given the way waves work. Imagine a wave moving through water at ten MPH. If you row toward it, from your point of view, the wave will appear to be moving faster. If you row in the same direction, the wave will appear to be moving slower. If you row in the same direction at ten MPH, the wave will appear to stay in the same place. Because the earth is moving, Michelson and Morley figured all they had to do was measure light waves as we, on the earth, moved toward or away from the sun or any other celestial source of light. The light waves would appear to move faster or slower, depending on which way the earth was moving, and they would have evidence of the aether.

Ah, but like yours truly trying to repair the bathroom faucet, what seemed at first to be a simple problem turned into a total disaster. No matter what these two did, they just couldn't get it right.[70] Regardless of whether they moved toward a source of light, away from it, upside down, backwards, or sideways, the speed of the light waves they measured never changed. They kept coming up with the same 186,282 miles per second every single time. All sorts of excuses were made: bad weather, defective equipment (including the equipment between their ears), you name it, all were cited.

The entire world of physics was in a quandary when our friend Al, who was still in his twenties, decided to weigh in on the subject. Between staring out the window and actually doing his job at the Patent Office, he quipped, There's a very logical reason why you can't find any evidence of the aether.... It doesn't exist. Outer space really is a vacuum just as Newton said. There's nothing out there to find. The scientific establishment scoffed. Undeterred, our friend Al went on to say: The results of your experiments are not wrong, they are right. The reason you always come up with the same answer for the speed of light-186,282 miles per second-is because *there is no variation in the shining of light.*[71]

Bingo!

From that moment, the world as we know it was changed forever. The reason is quite simple: with the possible exception of a few who took some of the biblical writers quoted in the previous chapter literally, the constant speed of light is the most fundamental understanding of the true nature of reality that has ever been revealed to mankind. What Einstein said was that no matter who you are, where you are, or how you are moving, light is the one thing (in fact, the *only* thing) in which there is no variation. You will always see it moving at 186,282 miles per second—period, end of story, no exceptions. Light is the universal constant. There is no variation in the "shining" of light (James 1:17 AMP).

Along comes this nerd who couldn't get into college and couldn't find a job when he finally did graduate, and, with one small equation, changed everything. Not only did he change everything we thought we knew about what we see when we look out the window, but he also included the entire universe: from the smallest particles imaginable to the farthest galaxies. In an instant, all notions of why things fall down instead of up, once again, had to be junked. Just when the scientific journey appeared to be truly getting to the bottom of things, once again, it was back to square one.

Nothing has changed, by the way. As we saw in earlier chapters, today scientists are earnestly searching for dark matter, dark energy, gravitons, and a "Higgs" particle, all of which they are so certain must exist that these, too, have already been named. The reality of their existence, like aether in its time, is almost taken for granted. Yet, as of this writing, no one has found any concrete evidence for the existence of any of these things. One or more of them may turn up someday, but if history has taught us anything, the odds are that some—if not all—of them won't.

CHAPTER 34

THE ONLY UNIVERSAL CONSTANT

"Only God Can"[72]

LET'S GO TO Zurich, Switzerland, (whoosh ... that was easy) at the beginning of the twentieth century, where we find our young friend Al goofing off at work. After scribbling some funny glasses on a picture of his boss, he writes down three letters and a number: $E=mc^2$. What this equation basically states is that mass (m) equals or is the same thing as energy (E). Just as water can be converted into steam, all mass can be converted into energy, and all energy can be converted into mass: material things. Like water and steam, they are just different forms of the same thing. In order to determine exactly how much energy (E) there is in any particular piece of something, you multiply that mass (m) by the speed of light (c) squared.

But how? What on earth or in the cosmos could possibly have given our friend Al the notion that energy and mass have anything in common? The answer is *c*. The *c* in the equation is short for "constant," and that constant is *light*. Al understood that "there is no variation" in the speed ("shining") of light throughout the entire universe. Instead of space and time being the same for everyone everywhere, Einstein said it is light that is the same for everyone no matter where you are or what you are doing. And as we have seen, the Bible says the same thing about

God. He is light (see 1 John 1:5) and He doesn't change (see Heb. 13:8, James 1:17, Mal. 3:6).

Because light is the universal constant, that means, as we first noticed in chapter eleven, that space and time are not the same for everyone everywhere. These things are actually perceived in different ways *relative* to a person's particular point of view. That is, time really will move at a different rate for people who are moving through space in different ways. In fact, even mass—objects themselves—will be perceived differently by people moving through space at different speeds.

Reality Check

> Give me a break.

At the speed we live our lives, most of these differences are not noticeable. However, they do exist. One hundred years of experimentation has left absolutely no doubt about it and (as we will see later) some very unusual things begin to happen once you strap on some engines and start to really move.

Al said that light was unlike anything in the universe. The fact that it is the one and only universal constant makes it the fundamental building block upon which the universe is built and by which it functions. His theory of relativity revealed that the universe was really not built through time inside of space; rather, the actual mainframe on which the universe is founded is Light. It is, in reality, *light*—electromagnetic energy—that is "upholding and maintaining and guiding and propelling the universe" (Heb. 1:3 AMP).

Until that moment, people considered the space around them—and the time it took to move through it—to be the unchanging constant container in which everything functions and exists. Ever since ancient cultures first began to discuss what have now become the various disciplines of modern science, the fact that a mile was a mile and an hour an hour no matter where you might be had never really been questioned—and for good reason. These apparently unchangeable things served people very well as a basis for understanding their world and relating to each other. Obviously—or certainly it seemed that way—anyone with two eyes and enough sense to open them could see that space and time were the same no matter where you were or what you were doing there. Who in their right mind would even think to consider that time and space were not the mainframe inside of which the universe unfolded and could be understood?

Enter our friend Al. What was he thinking? Well, he thought to ask a few questions. To start with he asked, "When you look at someone, what do you see?" Do you see the person, or do you see the light waves—photons—reflecting off that person, which your brain then interprets? When you look at *anything*—mass—do you see it or the light waves reflecting off it? When you listen to someone, what do you hear? Is it the person or the sound waves that have reached your ears?

If we, like Al, stop and think about it, light is how we relate to everything. Light is our reality. The absence of light is the absence of sight. The absence of light is the absence of heat. The absence of light is the absence of functioning atoms. The absence of light is the absence of electromagnetic energy. No light means no anything.

CHAPTER 35

THE FUNDAMENTAL THINGS APPLY

> It is necessary to remember that in the theory of relativity there is no absolute time. In other words, each observer has his own measure of time.[73]
> —Stephen Hawking

IN CHAPTER 33, we envisioned rowing a boat on waves that were traveling at ten MPH. If you row toward the waves, they will appear to be moving faster, while if you row away from the waves, they will appear to be moving slower. If you row in the same direction at ten MPH, the waves will seem to stay in the same position. Mr. Einstein said that this is absolutely true of every wave that exists except one: light waves. With light—and only with light—no matter how you move, there will be no variation in what you see.

Unlike anything else in the universe, if you move toward a source of light, you will still see it moving at 186,282 miles per second, just as you would if you were standing still. If you move away from the source of the light, you will still see the light moving at 186,282 miles per second. No matter how fast you are going in any direction, you will always see light waves traveling at 186,282 miles per second.

Let's explore this a bit by going back to chapter 11, where we left Fred on the moving boat and Ethel sitting still on the beach. We noted that when Fred bounced a ball on the ship, he saw it bounce straight back up and down. However, Ethel saw the ball move in a V pattern, because between the time Fred let the ball go and the time he caught it, the entire boat moved forward. Now let's set up the exact same scenario a few years later where we find Fred on a Space Shuttle heading to the moon. Instead of the beach, Ethel is lounging around at a stationary rest stop halfway to the moon where Fred is going to meet her for lunch.

As Fred pulls in, the Shuttle slows down to fifty MPH. Fred decides to throw a baseball out the window in the direction the Shuttle is moving at a target sixty feet, six inches away. He takes out the radar gun that he has packed in his luggage and measures the ball's speed at eighty MPH. Not to be outdone, Ethel pulls out her radar gun. Because the momentum of the Shuttle is added to the force of Fred's throw, and because there is no gravity or wind resistance in space, she measures the ball's speed at 130 MPH. She immediately decides to become a sports agent and call the Steinbrenners.

Now let's suppose that Fred decides to throw the ball from the back of the Shuttle in the opposite direction he is traveling. His radar gun measures the ball leaving at eighty MPH. From Ethel's stationary position at the rest stop—because the fifty MPH Fred is moving is subtracted from the force of the throw—Ethel now gets a reading of thirty MPH, and hangs up. When the two meet later, like most everything else they talk about, they disagree on what has actually taken place. For everybody except Fred and Ethel, it is easy to see why they are measuring different rates of speed. Ethel was standing still while Fred was moving.

Okay, here comes the only math we're going to do in this entire book. Speed equals distance divided by time. Let's say you travel one hundred miles in two hours. Taking the distance you traveled—one hundred miles—divided by the time it took you to get there—two hours—you find that your speed was fifty MPH. If you traveled 372,564 miles in two seconds, it would mean that your speed was 186,282 miles per second. That's it. The math portion of this book is over, and it's not cheating if you use a calculator. As I said, no algebra, long division, or pictures of ancient Greek guys drawing triangles in the sand. Speed equals distance divided by time. As we go on, the math is simple. It's the answers that take some getting used to.

Applying this to our two friends, when Fred threw the baseball in the direction he was moving, to Ethel the ball appeared to be moving at 130 MPH. However, if she had subtracted the speed Fred was moving—fifty MPH—from her equation, it would have accounted for the difference in their radar gun readings. They would both have come up with the same answer: eighty MPH. In the same way, when Fred threw the baseball in the opposite direction he was moving, if she added in the speed of the Shuttle—fifty MPH—to the thirty MPH reading on her radar gun they would again come up with the same answer: eighty MPH.

Now, instead of a ball, let's substitute the one thing—the only thing—in the universe that our friend Al said does not vary in the way it is measured no matter how one is moving. Yes, that is light. As we saw earlier, one of the things—among many—that Al showed the world was that light consists of tiny particles called photons.[74] This time let's say Fred throws some photons forward from the Space Shuttle. To do this, he unpacks his photon gun,[75] sticks it out the window, and pulls the trigger. Again, remember that there is no variation in the shining of light—it always appears to be traveling the same speed to everybody. What this means is that unlike the ball, there will be no difference in the speed that both Fred and Ethel measure the light traveling. Even though Fred is on the Shuttle moving toward the target when he shoots and Ethel isn't moving at all, each will measure the photon leaving the Shuttle at the exact same speed: 186,282 miles per second. If Fred shoots a photon from the back of the Shuttle, again, there will also be no variation in what both of them see. Both of their radar guns will read 186,282 miles per second.

Suppose the Space Shuttle doesn't slow down at all, and a very hungry Fred zooms past at 17,500 MPH. As he does, he shoots a photon out the window. Even though he is moving much faster, his speed will have absolutely no effect on the speed that either he or Ethel will measure the light traveling.

You can imagine what effect no gravity and no wind resistance would have on the speed of a baseball thrown from a spaceship moving at 17,500 MPH, but it will have absolutely no effect on light. Both Fred and Ethel will still measure the photon moving away from the Shuttle at 186,282 miles per second, regardless of what direction Fred shoots his photon. No matter what, *light always appears to everyone to be moving at 186,282 miles per second.* Again, as the writer James said, with "the shining of light, there is no variation" (James 1:17).

Reality Check

> Okay, you're doing great. If you feel like you may not understand this as well as you would like, that feeling is there not because the words are hard or misleading. The reason is that this just flies straight in the face of what we have come to believe is common sense. If we can remember that, in times past, facts long thought to be irrefutable certainties (a flat motionless earth at the center of everything) turned out to be anything but certain, it will help as we embark on the next leg of our trip.

Back at the rest stop, when Fred threw his ball out the window, Ethel either added or subtracted the speed of the Shuttle, and that compensated for the different readings on their radar guns. Ah, but now remember that our friend Al showed us that with light we cannot do that. Why? Because light is the one and only constant in the universe. Regardless of how fast or slow or in what direction we are moving, *we will always get the same answer for the speed of light:* 186,282 miles per second.

So, if it is true that everyone always sees light travel at the same speed—which they do—and speed always equals distance divided by time—which it does … what happens to our equation—speed equals distance divided by time—if one person is moving and the other is standing still while measuring the speed of light?

Reality Check

> You had to ask!

Remember that the reason Fred measured the ball moving at eighty MPH while Ethel measured it traveling at 130 MPH was because Fred was moving toward the target while the ball was in flight. That means he was closer to the target when the ball hit it. Unlike Ethel, he did not see the ball travel the same distance. They both saw the same ball in flight, but because they each had a different number for distance, they did not get the same answer for speed when they divided distance by the time it took for the ball to hit the target. They had to add or subtract the speed of the Shuttle to compensate.

Now, when Fred shined his photon from the moving Shuttle, both he and Ethel saw the same photon in flight and, as before, Fred was closer to the target when it hit. However, even though Fred saw the light travel a shorter distance, when they both divided the distance the light traveled by the time it took to get there, they both came up with 186,282 miles per second. The amazing thing is that they both got the correct answer. It has been proven that everybody *always* sees light travel at the same speed. There is no variation in its shining.

So, if the answer for the speed the light was traveling was correct for both Fred and Ethel but—because he was in the moving Shuttle—the distance Fred saw the light travel was different from what Ethel saw, there is only one possible thing that can be done to make the equation—speed equals distance divided by time—continue to be true under all circumstances—which it is. If Fred has a different number for distance, that leaves only one other part of the equation that can be altered if they both got the correct answer of 186,282 mile per second for speed, which they did.

Reality Check

> "You cannot be serious."[76]

I am completely serious when I say that the answer is *time*. As incredible as it may seem, both Fred and Ethel were perceiving time differently. Time was actually passing at a different rate for Fred in the Space Shuttle than it was for Ethel on the platform.

Reality Check

> Take a deep calming breath. The math is simple. It's the answers that take some getting used to.

Let's go back to the moving boat in chapter eleven. If Fred shoots a photon from the top of the cabin in a straight line down to the floor, off a mirror, and back up again while Ethel is watching from the shore, Ethel would see the light travel a farther distance than Fred because she was sitting still while Fred was moving in the boat along with the light. Like they each saw the ball, Fred would see the light go straight up and down while Ethel would see a V pattern. Ethel would see the light travel farther than Fred did.

But ... remember that because it is light they are looking at, they both come up with the same answer when they measure its speed. Like on the Shuttle, Fred saw the light travel a shorter distance than Ethel did sitting still. But, they both got the same correct answer of 186,282 miles per hour for speed. So ... if the distance is different but the answer for speed in both cases is correct, the only other possible way for the equation, speed equals distance divided by time, to be correct in both cases—which it is—is to change the value for time. Fred and Ethel actually understand that, because one is moving and the other is not, they truly are moving through time at a different rate. And, their marriage does not fall apart as both were previously sure it would.

Reality Check

> There is hope for us all.

In fact, not only does time bow to the consistency of light, but relativity also shows us that the perception of space itself will be different for the moving person compared to the stationary one. This is all to accommodate the fact that light is the one and only universal constant. Light never changes.

Okay, the answers certainly are hard to swallow, especially if, like me, the last time you heard a physics lecture was sometime in high school and your memory of it is, let's say, *nonexistent*. Like me, you probably had your mind on much more important things—the wonders of gender differences—at the time. If your mind was similarly preoccupied, you might respond to this as I did by saying, "This is crazier than science fiction."

Just imagine how all of this sounded to the staunch, egotistical, know-it-all scientific establishment when our friend Al first presented it one hundred years ago. Relativity—*time* changing to preserve the constancy of light—completely defies common sense.

In the decade that followed Al's initial papers on the subject, he remained a nut case in the eyes of almost every scientist who heard about it. However, as mentioned in chapter 19, all that abruptly changed in 1919. A British astrophysicist, Arthur Stanley Eddington, realized that during a solar eclipse, when the sun's light was blotted out for a few minutes, you could see distant stars that appeared behind the sun in the sky. If Einstein was right, the sun's gravity would shift the apparent positions of these stars compared to where they were normally seen when the sun was away from the line of sight to the stars. The findings proved that the sun's gravity bent the light from these stars exactly as relativity predicted it would. In just a few short years, our friend Al went from blubbering idiot to world-famous genius.[77]

As for the scientific community—along with everyone else on the planet—once again, our entire understanding of reality had to be scrapped. For who knows how many times and, doubtless not the last, it was "back to square one."

CHAPTER 36

THE EFFECTS OF RELATIVITY

> The only reason for time is so that
> everything doesn't happen at once.
> —Albert Einstein

BACK AT THE "original square," scientists couldn't believe the effect that $E=mc^2$ had on what seemed to be the simplest problems. For example, because space and time are intertwined—the "space time continuum"—what affects one affects the other. Because the sun's gravity really *does* bend space, it means that gravity also has an effect on the passage of time.

By the 1960s, atomic clocks had become so accurate they could distinguish billionths of a second. So scientists stuck one on top of a water tower where the force of gravity would be just a little bit weaker than it is on the ground—closer to the earth—where they placed another. Sure enough, the clocks ran at different speeds, confirming relativity. The stronger the force of gravity, the slower time passes. Global Positioning Satellites were inaccurate until scientists compensated for the difference between the rate that clocks tick in zero or micro gravity in space as compared to here on earth.

In 1971, some sixty-six years after Al first announced all of this, scientists attempted an experiment to confirm his prediction that time passes at a different rate if someone is moving as opposed to standing still. This time, they left one atomic clock at an airport and put another in a jet and flew it around the world at about six hundred MPH. When the jet got back, the clock on board had ticked slower than the stationary clock at the airport by exactly the amount Al said it would. This, again, was proof of the accuracy of relativity.

CHAPTER 37

THE INTERCHANGEABLE NATURE OF ENERGY AND MATTER

> Time is an illusion. Lunchtime doubly so.
> —Douglas Adams[78]

WHEN OUR FRIEND Al first came up with his theory, he wanted to call it "invariance theory" because its foundation is the invariant speed of light. However, the world did not share his perspective. People focused on the "relative" effects that the invariable speed of light had on everything else. Hence the name "relativity."

As we have seen, there is absolutely no variation in how light will appear to a person, no matter how the source of that light or anyone observing it is moving. 186,282 miles per second equals about 670 million miles per hour. Even if you are chasing after light at 600 million miles per hour, light will still appear to you to be traveling at 670 million miles per hour, exactly the same as it will appear to someone standing still. But what else will a person standing motionless see?

Suppose you are chasing after this light wave and your uncle is standing still watching you, scratching his head, saying, "Someday, I'm going to have to have a long talk with that boy." From your point of view, your speedometer would read 600 million miles an hour, and you would measure the light moving away from you at 670 million miles per hour. What would your uncle see? As we now know, he would also see the light moving at 670 million miles per hour, and, unfortunately for him, he would see you chasing after it at 600 million miles per hour. That means that, unlike what you are looking at, he would see the light moving away from you at 70 million miles an hour.

Back in chapter 7, I stated that we can relate to each other and do things together because we think we have the same perception of space and time. If I say, "Let's meet at the restaurant on the corner of First Street [length] and Second Avenue [width] on the second floor [height] for lunch in eight days at noon [time]," our plan will work because our senses tell each of us that these four dimensions are constant, or the same, for both of us. But Al showed us that this, actually, is not true. It is just that at the speeds and distances at which we live our lives, the differences in our perception of space and time are so small that meeting for lunch won't be a problem. A difference of a millionth of a second is not going to foul things up. However, the greater the rate in which we are moving differs, and the greater the distance between us differs, the greater the difference in our perception of space and time.

Suppose you decided to take a side trip to the nearest star—Proxima Centauri—at approximately 99.99999 percent the speed of light during the eight days leading up to our lunch date. The truth is that—let's take this slowly—if you made such a trip, I would be waiting about eighteen years for you to show up and your food would be a bit spoiled. However, from your own perspective, nothing would be out of the ordinary. That's right: from your perspective, you would not even be late, and our plans would still be on. At that speed in the rocket, about eight days would have gone by in your life and ticked by on your watch. However, when you got back, the world outside your window would be a very different place. Why? Because our relationship to space and time is not the constant that governs our reality. It is our relationship with light that never changes.

There is simply no getting around it: space and time are variable. The above example illustrates this in terms that emphasize the point, but in truth, it doesn't matter whether you are in a rocket, a train, or just walking by, whistling a happy tune. If someone else is moving at any speed different from your speed, he or she is moving through space and time at a rate all his or her own.[79] In fact, it turns out that not only will time pass at a different rate for someone moving at a speed different from your speed, but also objects in space (matter itself) will not appear the same to that other person.[80] In the case of light, however, the exact opposite is true. Light is the one and only thing in the universe that is invariable. It is constant and never changes. It always appears the same to everybody, no matter where they are or how fast they are moving.

Although it doesn't appear to be the case at first inspection, it is light that links us all together. It is light—not space and time—that is the fixed, unchanging frame of reference for everything in existence: all matter and all energy.

Yes, it all completely defies the common sense we have grown up with. But there is one unique thing about it. It is in amazing agreement with a book that's been around quite awhile.

As our friend Al showed us, energy and matter are interchangeable—they are two different forms of the same thing. He went on to show us that we can even know exactly how much energy (E) is in any mass (m) by multiplying it by the constant speed of light squared (c^2). And because the speed of light multiplied by itself is such a huge number, the answer is quite a lot. There is a "whole bunch" of potential energy in even a tiny bit of matter. For example, if you multiply the mass of a small amount of uranium or plutonium by 186,282 miles per second squared, you will get enough pure energy to blow up an entire city. This, in fact, is what occurred in 1945 when the first atomic bombs were dropped on Hiroshima and Nagasaki, Japan, bringing an end to the Second World War and launching the nuclear age.

Scientists have been converting matter into energy ever since. However, if mass can be turned into energy, the theory says that the opposite (turning energy into mass) should also be possible. The problem is that because just a few atoms (very small mass) create such an incredible amount of energy, it takes that same kind of massive force of energy to create just a few atomic particles. It took awhile for scientists to accomplish this, but it was finally done in 1997 at Stanford University in California when scientists focused billion-watt lasers in a vacuum on an area the size of a billionth of a centimeter. The transferring energy broke down the vacuum, creating an electromagnetic field that contained some matter and anti-matter. This creation of mass out of energy has since been accomplished numerous times at labs throughout the world.

The creation of mass out of energy is what the Big Bang theory says happened 13.7 billion years ago. This incredible force at the beginning of everything created the first atomic particles and formed stars. It then occurred again when the immense energy of supernovae exploding stars created most of the elements that comprise you, me, and the stuff at Wal-Mart.

CHAPTER 38

DOES ANYBODY REALLY KNOW WHAT TIME IT IS?

> For a thousand years in your sight are
> like a day that has just gone by.
>
> —Psalm 90:4

ISAAC NEWTON UNIFIED the moving tide, a falling glass, and the moon's orbit by showing that the same force—gravity—was responsible for all. In the same way, our friend Al unified space and time, which had always appeared to be two completely different things, by showing that the same force—light—was responsible for both. In the same way that the tides, things that fall, and the moon's orbit cannot happen without gravity, light ties together space and time. Scientists often call the fabric of our universe "space-time," because when you are talking about one you are also referring to the other. You can't take up space without taking time to be there. You can't meet your friend by picking just a time or a place. One won't work without the other.

Einstein also showed that the greater the difference in the speed you are going compared to someone else, the greater the difference in your relationship to space, time, and everything else in the universe, with just one exception: light. In the last chapter, we saw the extent of this difference when you took a side trip to a nearby star before our lunch date. To you, only a week passed. To me, however, eighteen years went by before you showed up ready to eat. Physicists call this phenomenon "time dilation." Though theoretically impossible, at least for a human being, Al pointed out that if you could actually travel at the speed of light, time would cease to exist.

Another way to think about time as a dimension (along with length, width, and depth) would be to say that just because you exist in one particular point in space, that doesn't mean it is not possible for you to be in any other. The same is true about time. Just as you exist in one out of an infinite possibility of spaces, you exist in one out of infinite possible moments of time. Just as it is your movement through space and relationship to gravity that determine what place you are in, it is also your movement through space and relationship to gravity that determine where you are in time. The term "space-time continuum" means that you change your relationship to time whenever you change your relationship to space. It is just that at the speed at which we live our lives the differences are not noticeable. If an astronaut spent two entire years in orbit, he would be less than half a second younger—compared to you and me—than he would have been had he stayed home.

We noted that Al's revelation about light came to him while he was imagining himself riding on a light wave around the universe. If you can imagine yourself doing that same thing—or riding on a photon (a light particle)—relativity means that the beginning of the universe, today's news, and the end of everything will all be happening right now. If you were light (a photon) or just riding on one, you would be the *alpha* (Greek for beginning) and *omega* (Greek for ending), the first and the last, all right now.

"I am the Alpha and the Omega, the Beginning and the End," says the Lord God, "He Who is and Who was and Who is to come" (Rev. 1:8 AMP).

Some years before the apostle John identified God as Light, his friend Peter wrote something quite interesting: "But do not let this one fact escape your notice, beloved, that with the Lord one day is like a thousand years, and a thousand years like one day" (2 Peter 3:8). Apparently, Peter was aware of time dilation some twenty centuries or so before anyone else knew about it.

We have seen that any difference in the speed people are moving will cause a difference in their relationship to time. When you took a side excursion to another star before our lunch date, because you were traveling at such an incredible rate of speed, there was a much greater difference between your relationship to time and mine. We can also describe this by saying that the greater the difference in our relationship to space, the greater the difference will be in our *perception* of time.

However, the difference between two people's relationship to space, which causes their different perceptions of time, does not necessarily have to be caused by speeding through space in a fast rocket. It can also be caused by the fact that there is a large amount of space between individuals. If your relationship to space is different from someone else's, then so will be your relationship to time. The only thing that will *never* change is our relationship to light.

Let's suppose that both you and I are barely moving at all, but you are on a planet in another galaxy about five billion light years from earth. That means it takes the light from that solar system five billion years to travel from there to here, and vice versa.[81] Suppose you asked your descendants to look through an incredibly powerful telescope at earth five billion years from now so that they could know what was going on here at the same time you had been alive there. Because of the vast amount of space between us, just a little movement on their part would produce very different results. If your descendants were standing still, they would see me. But, if they were in a car moving toward the earth, they might see the earth, say, fifty years in the future. If they were going in the opposite direction, they would see me spilling milk on Lisa Lipton in kindergarten.

So not only will the difference in what we perceive correspond to the difference in the rate we are traveling through space, but the difference in what we perceive will also increase as the difference in the space between us increases. Think about what you might see if you looked back at earth through that telescope and really hit the accelerator. Maybe you hopped in the Space Shuttle and started heading toward earth at a speed of 17,500 MPH, and then made a U-turn and began moving away from the earth. As you accelerated, slowed down to turn, and then sped up in the opposite direction, almost any time—past, present, or future—in the history of this planet could appear to you to be happening right then.

In fact, for anyone with the ability to move freely through the universe at any speed he or she chooses, all the events in space and time will be within the reality of the present moment.

"I am the Alpha and the Omega, the first and the last, the beginning and the end" (Rev. 22:13).

Although the how and why of it is beyond the scope of our discussion here, not only does time pass differently for one person if he or she is relating to space differently from someone else, but also objects themselves (mass) are actually a different size than they are when they are at rest. As with time dilation, the difference is only noticeable when speeds get really high. One of the things this means is that whatever you are measuring—time, space, mass—it won't be exactly the same for you as it will be for another person, simply because the two of you are different people moving at different speeds at different spots in the universe. It's all "relative" to your point of view.

Earlier, I mentioned what seems the unnecessary antagonism that exists in our society between what we have labeled secular science and theology (or "Creation Science"). In the seventeenth century, the argument was over whether the sun went around the earth or the earth went around the sun. Because our perception of space and time is different depending upon our relationship to it, today's argument over how old the earth is may also miss the point.

How old would the universe appear if we were looking at it from a different planet with a different mass in a different galaxy in a different orbit? Right where we are at this moment, we know that we are spinning around each day on the earth's axis at a speed of about 500 MPH.[82] Meanwhile, our planet is taking one year to complete an orbit around the sun at a speed of no less than 66,600 MPH. That is fast—more than 1,100 miles per minute—not to mention the fact that during that same minute, we also traveled 7,500 miles on our trip (along with the rest of the entire solar system) around the center of the Milky Way galaxy.[83] And last, but certainly not least, we cannot forget that during that same *minute,* you (and the entire Milky Way) just skipped 11,100 miles down the road toward or away from a galactic neighbor. Let's see, that adds up to movement through space of more than 19,000 miles, in one minute, in four different directions ... all at once.

Reality Check

> I'm dizzy.

If our perception of time is different depending on where we are in space and how we are moving through it, and if time also moves faster or slower depending on the amount of gravity at any particular location,[84] how old does anything appear to be from any other place in the universe? Knowing that space and time are relative, why are we still using space (it all started in a little ball and exploded outward) and time (some 13.7 billion years ago) as the constants by which we measure everything? How old would the universe be if we were moving a little slower? What about if we were going faster? What would the answers be if our planet were in another galaxy?

As we go on to the next chapter, we find the Bible defining God as light. We find relativity showing us that light is the only constant in the universe and at the speed it travels, time ceases to exist. Then we find the Bible repeatedly saying that God is someone Who is in all times at the same time (see Isa. 4:4; Rev. 1:4, 8; 4:8; 21:6; 22:13).

CHAPTER 39

THE PARADOX OF LIGHT

> The truth may be puzzling. It may take some work
> to grapple with. It may be counterintuitive. It may
> contradict deeply held prejudices. It may not be
> consonant with what we desperately want to be true.
> But our preferences do not determine what's true.
> —Carl Sagan

AFTER OUR FRIEND Al revealed relativity to the world in the early 1900s, all the painstaking scientific research of the prior century came crashing down. There was nothing for scientists, now left sitting in the middle of a mess scratching their heads, to do but start all over again. So that is exactly what they did.[85]

Back again at the original square, scientists set up the same double-slit experiment Thomas Young had started with back in 1800. We discussed this in chapter 31. Remember that the light Young sent through the two slits in one screen produced a wave pattern on the second screen. But also remember that a century later Einstein had produced evidence that light was not a wave but a particle as Newton had surmised. Then, displaying the kind of common sense that occasionally reassures us all, researchers said, "Hold everything, not so fast," and went about constructing a machine capable of detecting light particles. They set the machine up behind one of the slits between the first screen and the second screen. Obviously, this would quickly put an end to this silly contradiction and conclusively show that light was either a particle or a wave.

To begin with, researchers went into a nice big room and performed the experiment the exact way Young had more than one hundred years earlier. When they did—what a surprise—they got the same result. They saw the same wave pattern on the second screen that Young had seen. However—and it's one big *however*—when they turned on the particle detector and shined the light through the same two slits they had shone it through just a few minutes earlier, they found something *very different* from what they expected.

Reality Check

Remembering that truth really is stranger than fiction, let's just relax and take this next part leisurely.

What was it they saw on the second screen? What they saw was—and please read this slowly but surely—*light behaving like a particle.*

Reality Check

See, that wasn't so bad.

Okay, let's go over this again. When the scientists performed the experiment the same way Young had done it, just like Young, they saw a *wave pattern* on the second screen. But—and it's also a big *but*— a few minutes later when they turned on the particle detector between the two screens and performed the experiment exactly as before, they detected *particles* of light going through the slit in the first screen where they had set up the particle detector. Then when they looked at the second screen, instead of a wave pattern, what they saw was a whole bunch of tiny little spots individually building up behind *both* slits. Imagine dropping grains of sand through two tiny holes and the haphazard pattern you would see when the grains landed underneath. That was the pattern of dots scientists saw on the second screen. Clearly, this showed that light was made of particles. There was no hint of a wave pattern.

So where does that leave us? The scientists had performed the experiment just as Young had first performed it and then performed it a few minutes later just as they had performed it the first time, except for *only one* difference. The second time they turned on the particle detector.

"Okay," you say. "What's the catch?" That is a good question, because that is exactly what every single scientist wondered as well. Just a few minutes before, with the particle detector turned off, a wave pattern had appeared on the second screen, clearly showing that light was a wave. But when they turned on the particle detector, it detected particles, and they saw a pattern on the second screen that indicated light was made of particles. That being the case, what do you suppose the scientists did next? That's right—you're getting the hang of this—they turned the particle detector *off*.

When they did, what do you think they saw? They saw light behaving like a wave. So—you guessed it, and you are doing very well—they turned the detector back *on* again, then *off* again, then *on* again, then *off* again, then *on* again ...

Each and *every* single time the scientists turned on the particle detector, they detected particles and saw a random area of dots build up on the screen behind the two holes. Each and *every* single time they turned the detector off, they saw the light producing a wave pattern at the exact same place on the screen.

Reality Check

> If, despite everything you have just read, you continue to say, "Hold everything, I must be missing something; that is not possible, it has to be one or the other," let me congradulate you because in a matter of just a few seconds you were able to conclude what came from the mouths of what are considered to be the most brilliant scientists on the planet. They said exactly what you said: "That is ridiculous. A

> particle is a tangible, physical thing. A wave is not a thing. A wave is energy; it travels through things like water. A thing isn't a wave, and a wave isn't a thing."… In fact, right now you know the essence of what has been written in all the scientific literature on this subject during the past eighty years. So, because you now have a thorough and complete understanding of the quantum dilemma, I see no reason why you shouldn't apply for a federal research grant. Your guess is certainly as good as anybody else's.

Faced with this riddle, it didn't take long for researchers to develop a machine that could shoot only one light particle—one photon—at a time at the two tiny holes.[86] It seemed obvious that, regardless of whether the particle detector was on or off, a single photon shot at the screen toward the two slits could only go through one hole or the other and produce a dot on the second screen. So, they set up the particle detector behind one of the holes and shot photons at the first screen in the general direction of both holes, one at a time. As expected, when the particle detector was turned on, about half of the time it detected a particle going through the hole it was behind, and a corresponding area of tiny dots appeared on the second screen behind both holes. What this means is that as long as the detector was turned on, a particle went through one slit or the other.

However (it's that big *however* again), when the scientists turned the particle detector off, even though they shot only one photon at a time in the direction of the two holes, they still saw a wave interference pattern build up on the second screen behind both holes.

Yes, your initial reaction to that statement is correct. That is impossible, at least according to any concept of reality as we know it. The results of the experiment make as much sense as dropping one stone in a lake, waiting a minute or two, and then dropping another stone and seeing the waves from each stone interfere with each other the way they would if both stones hit the water near each other at the same time.

But the truth of the matter had absolutely no respect for anyone's ideas, beliefs, or opinions. In fact, it had absolutely no regard for any known concept of reality or all the research and work that had gone into establishing those concepts throughout the entire history of mankind.

You see, despite shooting only a single photon at the screen, with the particle detector turned off, the wave interference pattern continued to appear on the second screen every time. Even though only one photon was shot from the gun, there was no evidence of a single photon going through either one of the holes and hitting the second screen. As long as the detector was turned off, all that was seen was a wave pattern.

As you might imagine, since this experiment was first performed, scientists have racked their brains looking for an explanation. The most logical answer seemed to be that the photon somehow divides on its way to the screen and goes through both holes, producing a wave pattern. However, throughout the countless times this experiment has been reproduced during the past half century, no evidence has ever been found that the photon divides and goes through both holes.

Today, it is accepted by almost all of the scientific community that the reason light behaves like a wave when the particle detector is turned off and like it is made of particles when the detector is turned on is simply because light is both a wave and a particle. That's it! It's as simple as that. That's all there is to it. Light exhibits both wave and particle properties. As with the Uncertainty Principle, the words are in plain English. What makes it difficult to comprehend is that it simply makes no sense when we try to fit it into the reality we have taken for granted all our lives.

Because it does defy the basic tenets of science and common sense that we have never even thought to question, as you might imagine, there are those who think that there has to be some fact or aspect of the experiment that has been missed. *However*, nothing has been missed. Remember, this experiment was first performed more than eighty years ago. Since then, the incredible number of attempts to find some flaw or make some kind of logical sense out of it cannot be counted. It has literally been performed hundreds of thousands—if not millions—of times with innumerable variations. Despite some of the most desperate hopes to the contrary, the result has never changed. Like a globe that no one falls off of, the results simply ignore people's most basic concepts of reality. Light is both a wave and a particle.

CHAPTER 40

STANDARD PHYSICS

Here's another fine mess you've gotten us into.
—Oliver Hardy

AS I MENTIONED, this "double slit" experiment is now standard in most college physics labs. Students are required to learn that the dual nature of light is a basic aspect of reality. What they are not asked to do is make any sense out of it. Who can? It is, literally, impossible to believe, except for one nagging detail. No matter what scientists do, the results never change. To this date, the results have been totally impervious to all attempts to interject even a little common sense.

In fact, instead of finding a mistake with the experiment or coming up with a logical explanation for why the findings are this way, what the past eighty years of intense effort have proven is that *not only do light particles—photons and electrons—behave this way, but so do all atomic particles!* That is, *atoms*, the building blocks of everything in the universe, *behave in the same manner.*

Reality Check

> The fact that all matter has "wave-like" properties was first discovered in the 1920s by Louis de Broglie.

Science explains the dual nature of atoms, photons, and all atomic particles by saying that, to begin with, they exist as a wave, or to be more precise, scientists say they have a "wave-like" nature. That means that everything—including you and me—has a wave-like, rhythmic nature.[87]

Applying this to the double-slit experiment, we have a single photon coming out of the gun in a "wave-like" state. Because it is not yet a particle, that means it can go through both holes at the same time.

Reality Check

> Wait a second.
> This is like saying a single thing can be in two places at once.

> No, it's not *like* saying that. It's saying exactly that.

Within this wave-like state, there exist certain possibilities. The photon could retain its wave nature, or it could take on the nature of a solid object—particle—in which case it will only be able to go through one hole at a time.

As I said at the beginning of the chapter, for the scientific community, it was and still is a big mess. In fact, it is probably the biggest mess in the history of science.

Okay, so why is it the biggest mess in the history of science?

As we are sitting here looking at particles at one moment and waves the next, there is this uncomfortable and rather annoying question that just won't go away. In the same way that it came to your mind as we were discussing this, scientists could not avoid the question: *What is it that determines whether light will be a particle or have a wave nature?*

Reality Check

> Okay, let's take this slowly.

We do.

Reality Check

> I realize you saw this from the start. The problem was believing it. The answers take some getting used to.

But *it is true*. In the experiment, when the particle detectors were turned on, a particle was detected. When they were turned off, a wave pattern appeared on the second screen. Who decided when the detector was on or off and, therefore, what appeared on the screen? It was the scientists themselves. When they decided to look for a particle and turned on the detector, that's what they found. When they turned it off, they detected a wave. It was people who determined whether or not the wave-like nature of the photon became a particle.

As noted, this experiment has been performed countless times in every lab, college, and university throughout the world for more than eighty years, and the results have never changed. Not only photons, but also all atomic particles—everything that makes up all matter—behave in this same manner. What this experiment has proven is that, as conscious observers, we have the ability to influence the nature of our reality.

Reality Check

> "[Jesus] said, 'Because of your faith, it will happen.'"
> —Matt. 9:29 NLT
>
> "Now faith is the substance of things hoped for, the evidence of things not seen."
> —Heb. 11:1 KJV

This quantum paradox of an observer-created reality is nothing new. What's new is the grudging acceptance of it by the scientific community. One of the most famous and well-documented arguments on the subject occurred between our friend Al and Niels Bohr. It began at a scientific conference in Belgium in 1927 and continued throughout their lifetimes in public as well as in private.

Al said, "I like to think that the moon is there even if I'm not looking at it." He reasoned that if single atoms existed in this "semi" real state until someone decided to look for them, because the moon is made of atoms, obviously, this would apply to the moon as well. As hard as this may seem to argue against, Bohr did so, and he did it successfully. Regardless of how big or small something might be, when you get right down to it, the atoms it is made of exist only in a potential state until an observer activates their potential. The truth, it seems, continued its policy of having absolutely no regard for anyone's—even Einstein's—fundamental concepts of reality.

Reality Check

> "[God] speaks of the nonexistent things ... as if they [already] existed."
>
> —Rom. 4:17 AMP

When you pause and consider the fact that everything—including you and me—is made up of atoms and that no flaw can be found in the double-slit experiment, you can see how it truly has produced one "fine mess" in the world of science. As the twentieth century progressed and the brilliant men and women of the scientific community used quantum physics to technologically transform our planet, many of them—and their universities—all but ignored this illogical state of affairs because, regardless of whether or not it made sense, *it worked*. However, raising the possible implications of what quantum physics might actually mean in a larger sense was seriously frowned upon—it could even ruin one's career. Through the twentieth century, most people held on to the hope that someone would find a way to make sense of it all. Only now, after more than eighty years, is this observer-dependent nature of reality being accepted by the scientific community as a fact that simply cannot be denied.

It is interesting to note that at one end of the halls of academic freedom exist the quantum physicists who teach that matter exists in a potential state until a conscious observer directly looks for it, while, at the other end of the building, the evolutionists teach the opposite: matter came first and organized itself into consciousness.

NO EXPLANATION

> If quantum mechanics hasn't profoundly shocked you, you haven't understood it yet.
> —Niels Bohr

I UNDERSTAND THAT what I have said in the two previous chapters is so counterintuitive (that is college professor talk for "crazy") that even though there is not a single word that is not a common part of everyday language, to absorb it all in just a few minutes is not easy. As I've said before, there are many scientists who still find it hard to accept. Driven by the belief that there has to be a logical explanation consistent with reality as they see it, some physicists continue to insist that the since a wave pattern appears on the second screen even though just one photon at a time is shot at the first screen, the photon somehow divides and goes through both holes. [88]

At this point, you might ask (as I did), "If this is truly such a basic part of our lives, why have I never heard of it?" The reason is quite simple. Scientists are in the business of explaining things, but no one can explain this. It is hard to make it in any profession by going, "Duh, beats me." Science simply has no explanation. So before our brains short-circuit and our hair starts to stick straight up in the air, let's consider something else.

This happens to be what the Bible has always maintained is the true nature of this physical reality.

"By faith we understand that the worlds were prepared by the word of God, so that what is seen was not made out of things which are visible" (Heb. 11:3).

CHAPTER 42

CAUSE AND EFFECT

> It is far better to grasp the universe as it really is than to
> persist in delusion, however satisfying and reassuring.
> —Carl Sagan

WE HAVE SEEN that at the foundation of quantum physics lies the inescapable conclusion that reality at the atomic level (not forgetting that everything is made of atoms) is dependent on the observer. Looking back two thousand years, we find that a man named Jesus said the same thing.

The people of that time were not much different from us. This idea was as crazy to them as it is to most of us. So Jesus illustrated the concept in a number of ways.

At times He stated it plainly, as in Matthew 9:29 when He said, "It shall be done to you according to your faith." Faith, of course, is a belief in something for which, at the time, there is no concrete evidence. Having faith carries with it the presumption that a time will come when what we have originated in our imaginations will become an actual fact. As we see in this passage, Jesus said that this is the source of what happens to each one of us. Not only that, He did not set any limits. In Mark 9:23, He said, "All things are possible to him who believes." To put it another way, one might say, "All things are possible if a person believes that what he imagines will, in fact, become a real occurrence in this physical reality."

Remember our discussion in chapter six of the power of imagination and the role it plays in human creativity? Everything made by human beings began as an idea—as a picture in someone's imagination. Quantum physics has shown us that this is the true nature of atomic (all) reality. Jesus plainly stated that the behavior of things made of atoms is dependent on the observer.

> For truly I say to you, if you have faith the size of a mustard seed, you will say to this mountain, "Move from here to there," and it will move; and nothing will be impossible to you.
> —Matt. 17:20

> If you had faith no bigger than a tiny mustard seed, you could tell this mulberry tree to pull itself up, roots and all, and to plant itself in the ocean. And it would!
> —Luke 17:6 CEV

> For verily I say unto you, That whosoever shall say unto this mountain, Be thou removed, and be thou cast into the sea; and shall not doubt in his heart, but shall believe that those things which he saith shall come to pass; he shall have whatsoever he saith.
> —Mark 11:23 KJV

Let's go back to the way scientists searched for the truth in the double-slit experiment. If they did nothing, they detected light as a wave. However, when they built a particle detector and turned it on, they found what they were looking for.

The same was true in Jesus' day. Jesus was called "rabbi" or teacher. He taught with words, but more so by example. The examples were the people who approached him looking for something. Like the scientists who turned on the particle detectors, when they acted with the intention of finding it, they found what they were looking for. Others who were nearby but not looking for the same thing did not find it, and had no one said anything, they would have never known that anyone did find it. Here's a story from Luke.

In the crowd was a woman who had been bleeding for twelve years. She had spent everything she had on doctors, but none of them could make her well. As soon as she came up behind Jesus and barely touched his clothes, her bleeding stopped.

"Who touched me?" Jesus asked.

While everyone was denying it, Peter said, "Master, people are crowding all around and pushing you from every side."[89]

But Jesus answered, "Someone touched me, because I felt power[90] going out from me."

The woman knew that she could not hide, so she came trembling and knelt down in front of Jesus. She told everyone why she had touched him and that she had been healed right away.

Jesus said to the woman, *"You are now well because of your faith.* May God give you peace!"

—Luke 8:43-48 CEV, emphasis added

In another instance:

When Jesus was going into the town of Capernaum, an army officer came up to him and said, "Lord, my servant is at home in such terrible pain that he can't even move."

"I will go and heal him," Jesus replied.

But the officer said, "Lord … just give the order, and my servant will get well. I have officers who give orders to me, and I have soldiers who take orders from me. I can say to one of them, 'Go!' and he goes. I can say to another, 'Come!' and he comes. I can say to my servant, 'do this!' and he will do it."

When Jesus heard this, *he was so surprised* that he turned and said to the crowd following him, "I tell you that in all of Israel I've never found anyone with this much faith!" …

> Then Jesus said to the officer, "You may go home now. Your *faith has made it happen.*"
> —Matt. 8: 5-10, 13 CEV, emphasis added

Many times, the people Jesus encountered asked Him to come, pray, and touch the person who was afflicted. Most found it easier to believe (have faith) that they would find what they were looking for when Jesus was close to the individual. But, as demonstrated in the stories above, it was not necessarily how close Jesus was or what He did, but the person's own view of reality—faith—that enabled the healing.

The Gospels record the stories of a number of people who were blind and received sight. In one instance, Jesus spit in the man's eye, laid His hands on him, and after the man's partial sight was restored, He put His hands on him again (see Mark 8:22-25). In another instance, Jesus spit on some dirt, made clay out of it, and put it on the man's eyes (see John 9:1-7). In yet another situation, He simply said, "Go, your faith has made you well," without touching the person at all (see Mark 10:46-52).

In the first two cases, His actions gave each man a reason to believe. In the third, such action was not necessary because the person already possessed what was needed for the present reality to change: faith. Jesus said, "Seek, and you will find; knock, and it will be opened to you" (Matt. 7:7). People who were not seeking healing from Him were not healed. They determined their own reality. This is illustrated by what occurred when Jesus went to His hometown: "He did not do many miracles there because of their unbelief"[91] (Matt. 13:54).

We see an interesting example in which faith was about to create reality in the negative sense in the first chapter of Luke. In this instance, Zacharias, the father of John the Baptist, was made *"silent and unable to speak until the day when these things take place* [his son was born]*, because you did not believe my words …* [so that God's words could] *be fulfilled in their proper time"* (Luke 1:20, emphasis added). Here we see God preventing Zacharias from creating a reality that would mess up His plans. The principle in science and in the Bible is that we find what we are looking for.[92] Our reality begins in *us*.

CHAPTER 43

AN OBSERVER-CREATED REALITY

> We are all agreed that your theory is crazy. The question
> which divides us is whether it is crazy enough to
> have a chance of being correct.
>
> —Niels Bohr

PARIS WAS LOVELY in the summer of 1982, as it always is. While most people were outside enjoying the city's beauty, inside an innocuous laboratory something was occurring that—though little noticed at the time—may well cause a more profound redesign of mankind's picture of his environment than anything Newton, Magellan, or Galileo ever did. It was there that a physicist named Alain Aspect answered a question that had been raised by one of our friend Al's former students, a physicist by the name of John Wheeler.[93]

    Let's go back to the room in chapter 39 where the scientists set up the photon (light) gun and the particle detector along one of the paths behind the two holes in the first screen. Remember that they shot only *one photon* at a time from the gun. With the particle detector turned on, about half the time they detected a light particle traveling along the path where the detector was set up and a haphazard area of single dots built up behind both holes on the screen. With the particle detector turned off, the scientists saw a wave pattern behind both holes.

This experiment has been conducted thousands—if not millions—of times for nearly a century, and the results have never changed. For this reason, most (if not all) quantum physicists now accept as fact that the outcome of the experiment depends on whether or not the observer decides to turn on the particle detector. Furthermore, all atomic particles—atoms and the things of which they are made—behave in this manner.

Observing all of this back in 1979, Mr. Wheeler engaged in a highly suspicious activity. He went to thinking … objectively.[94] This caused many to wish he had just gone home and watched *Fantasy Island*.[95]

The first thing he wondered about was what we first noticed back in chapter 39. Even though the particle detector was set up along just one of the paths, the dots appeared behind *both* holes. That indicates that even though half the time the particle detector did not actually detect a particle, the fact that it was turned on was enough to affect what occurred along both paths. Turning on the detector was enough to give the photon the characteristic of a particle. Our friend Al called this sort of entanglement "spooky," because there was no better scientific explanation for it.

Mr. Wheeler wondered about the distance of this "spooky" interaction. What would happen if he set up the particle detector way down the line? What if he waited until after the light had passed the first screen before he turned it on? Certainly, once the light had passed the first screen, turning a particle detector on or off could not have any effect on what was seen on the second screen. The second screen was only there to show what happened at the first. Once the light passed through the first screen, it meant that it had either gone through both holes as a wave or through one of the holes as a particle. Right? Clearly, that was obvious.

Was it?

Mr. Wheeler wasn't so sure. This led him to do something that Galileo and many others throughout history had proven to be quite dangerous. He threw out his preconceived conceptions of reality. He ignored any assumptions he had as to where the data had to fit and also refused to ignore any questions the results raised simply because the answer would not conform to the present unquestioned reality.

As we have seen, the amount of peril involved in this type of behavior is proportional to the amount of time the current "truth" has been an accepted certainty. When Galileo ignored the reality of his time in favor of the facts before his eyes—an earth-centered universe had never really been questioned since man first gazed at the stars—his actions proved to be nearly fatal.[96] For some of his contemporaries, it *was* fatal. Wheeler's query looked for a similar hole in reality's foundation.

Here is what he wondered. If this "crazy" notion of an observer-created reality was actually true, wouldn't the answer to what occurred at the first screen have to wait until somebody turned on the particle detector—regardless of when that person made the decision to do so? If what happened at the first screen really did depend on what a conscious person did or did not do—turn on the particle detector or leave it off—wouldn't that remain true even if the light had already passed through the first screen?

Reality Check

Uh-oh.

Now you know why some thought that Mr. Wheeler actually belonged on *Fantasy Island*—"De plane boss, you missed de plane!" You see, if what he was saying actually turned out to be true, it would mean that the constraints of time and space would be more illusion than real because *reality itself would have to wait for someone to create it*. The conscious observer would actually be creating the reality of the past!

Reality Check

This is so "totally weird" that we need to stop for a few seconds to process it. The experiment and the result are simple. It is the possible answer that hits a neural roadblock (a scientific way of saying, "*no way!*"). Young had devised the whole experiment so that the second screen would show what had happened at the first—either a particle went through one of the holes or a wave went through both. However, for nearly eighty years now, the results have continued to show that it is the observer who decides which of the two will happen.

What if the particle detector was miles away from the first screen and the second screen was farther still? What if the person conducting the experiment did not decide whether to or not to turn the detector on until after the light had already passed the first screen but before it got to the second? If this observer-created reality idea was true, it would mean that what happened at the first screen would be determined after the light had already passed it. It was not just the reality of the present or future that would be dependent on the actions of the observer, but also *the reality of the past*.

Theoretically, Wheeler observed, if the results really *did* depend on the observer, there would be no time limit to this principle. The reality of what happened before any observers came along would simply have to wait for them to create it.

To get a better picture of how this might be tested, let's step away from the room with the two screens and go outside—way outside, like outside the solar system. Suppose that instead of a split second between the time the light passed the first screen and hit the second, you had a lot more time to work with—say one hundred years or so. Suppose that you set up a mirror behind one of the holes in the first screen that reflected the light going through it in the opposite direction of the light going through the other hole. In other words, if the light went through the hole on the right, it would continue to go due east, while if it went through the hole on the left, it would hit the mirror and go due west. Remember that in the experiment, if no particle detector is turned on, a wave pattern appears on the second screen behind both holes, indicating that a wave went through both holes after shooting just one photon at a time. In our case, this means that without the particle detector turned on, after you shoot light at the screen, you will have one light *wave* heading through space at 186,282 miles per second to the east and another light *wave*, after it hit the mirror, heading just as fast in the opposite direction.

Let's suppose that we have two more screens set up at the end of each path. The first is set up on another planet one hundred light years away to the east, and the other is set up one hundred light years to the west.[97] So according to the experiment, as long as no particle detector is turned on, you will have two waves traveling in opposite directions, and a wave pattern will show up on each of the screens one hundred years from now.

That's all just fine … unless …

"Unless what?"

Unless ninety-nine years from now, you have nothing better to do but click on a particle detector.

Yes, there you are, out in space near the end of one of the paths and, just to drive most of academia crazy, you decide to turn on your particle detector. Surely that cannot have any effect on what happened at the first screen ninety-nine years ago? Clearly, what happened at the first screen can only be a distant memory. Of course that is true … unless … Unless what?

Unless the outcome really does depend on a conscious observer.

This is where Professor Aspect and the beautiful city of Paris enter the scene. It was there in 1982 that he and his colleagues developed the laser technology necessary to actually physically test and answer this question with enough precision to settle the issue.

You guessed it.[98] What Professor Aspect found is that it does not matter how long you wait. If you turn on the particle detector along one of the paths a split second, a day, a week, a month or ninety-nine years after the light passes the first screen, *it makes no difference.* The result does not change. As long as you turn on the detector before the light reaches the second screen, the answer is the same. If you detect a particle, you get a dot on your screen. If you don't detect a particle on your screen, because you turned on the detector, a dot will appear on the other screen two hundred light years away. Regardless of how much time and space exists between the two screens, if you turn on the detector, the light will exhibit the characteristics of a particle. *The observer does decide the outcome.*

Okay, that wasn't really too bad. You have two waves going in opposite directions for ninety-nine years. If you do nothing, a wave pattern appears on both screens. If you decide to turn on your particle detector, either it detects a particle, or if not, a particle appears on the other screen two hundred light years away. At least we have that much straight.

Yes, but hold on a second.

Reality Check

> I'm not sure I want to.

What about Mr. Wheeler's original question?

Remember that turning on a particle detector affects what happens along both paths. So if you turn it on and don't detect anything, how does the light heading toward the other screen, 200 light years away, know that a dot should appear? Or if you turn on the detector and do detect a particle, how does the light heading toward the other screen, two hundred light years away, know that nothing should appear?

Reality Check

> Again, the question is not hard.
> It's the answer that takes some getting used to.

Science tells us that for ninety-nine years, you had two "virtual" or possible particles traveling through space. Whether or not one or the other actually becomes a real tangible particle (material thing) remains up in the air until you decide whether or not to turn on your detector.

It's up to you.

Let's look at it from the perspective of the waves. We know that if no one turns on a detector, a wave pattern appears on both screens. Let's imagine the two waves traveling in opposite directions for ninety-nine years and ask the same question. What happened to the waves that for ninety-nine years had been traveling in opposite directions and are now 198 light years away from each other when you decide to turn on your detector?

Here's the scientific explanation: "*Poof!*"[99]

Once you turn on a particle detector along one of the paths, the wave that was traveling along the other path suddenly disappears.

Does this mean it instantly vanishes?

Reality Check

> Haven't you had things like socks, money, or cell phones instantly vanish? There you go.

The actual fact is that space and time are not the impenetrable barriers that we, for all of human history, have thought they were. It doesn't matter how far or how long; it is all in the hands of the conscious observer. Up until this point, physics, science, and reality as we have known it required two objects to be close enough to each other to allow the energy from one to take some time to travel through space, reach, and thereby affect the other. However, the knowledge of quantum leaps and the research carried out by Mr. Aspect and others have now shown that this space-time limitation does not really exist. Light does not conform to the limits of space and time. It's the other way around. Light exhibits the characteristics the Bible has always maintained it has. The Bible gives God the physical characteristic of light and says He exists in all space and all time at the same time (see chapter 32). Further studies have only served to confirm Mr. Aspect's findings. At this writing, the latest took place in Paris in 2006.[100]

The whole thing sounds crazy! As quoted at the beginning of the chapter, according to Niels Bohr, whom history records as one of the greatest scientists who ever lived, that just happens to be the one thing that gives it a chance of actually being correct. Richard Feynman, quoted at the beginning of chapter 29 and considered by many to be one of the greatest physicists who ever lived, had an impossible time comprehending it. After a series of lectures and filling numerous chalkboards with countless numbers and symbols that only a few other people on the planet could even begin to make sense of, he paused and, after pondering it all, summed it up this way: "I think I can safely say that no one understands quantum mechanics."

CHAPTER 44

VIRTUAL PARTICLES AND PROBABILITY WAVES

> We often stumble onto the truth, but most of us brush ourselves off and pretend it didn't happen.
> —Winston Churchill

WHAT HAPPENS TO one particle instantly affects another particle or wave regardless of how much space and time exists between the two. Physicists describe this by saying that reality is "non local." The other way they describe this phenomenon is to say that the particles are "entangled," meaning they somehow appear to remain connected beyond the boundaries of space and time.

There are some other common terms that pop up when physicists describe "entanglement" or "non-locality." Going back to Mr. Aspect's findings, we know that we will have a wave traveling through the universe for ninety-nine years unless someone decides to turn on a particle detector. If that happens, the wave that we thought was heading our way ... suddenly isn't. Science calls these waves that in the future *might* not end up to really be waves "probability waves," or a "possible wave function." This means that the light is functioning as a wave on both paths, but when you turn on the detector and discover a particle along just one of the paths, the wave on the other path collapses. It simply ceases to exist.

Another term is a "virtual wave." And because in this instance we are turning on the detector way down the line, which means that we will detect a particle or one will hit the other screen, what we also have during the ninety-nine years before the on-switch is thrown is a "virtual particle."

Until a conscious observer looking for one or the other decides which he or she is looking for, what we have for ninety-nine years is an indefinite probability or possibility that there could be a wave or a particle. And those probabilities or possibilities will remain just that until a conscious observer enters the picture and decides the outcome. It makes you wonder if these scientists are really mutants who messed up their DNA by playing with too many chemicals after class.

Reality Check

> I will admit that for most of my life I would have agreed with this assessment. However, because I have been married for some time now and have two teenage children, having my basic concepts of reality turned upside down has become pretty much a daily occurrence. So I was soon able to accept this reality as true and not let my own opinion that the whole thing is utterly ridiculous complicate matters.

There have been other interesting variations of Mr. Aspect's experiment. In one, scientists put what amounted to polarized sunglasses along one of the paths. The sunglasses changed the polarity of the photon going down that path, making it distinguishable from the photon going down the other path. In this case, even without a particle detector, two random areas of dots appeared on the screen.

Next the scientists set up another pair of sunglasses along the same path. The second pair reversed the polarization of the photon that went through the first pair back to what it originally had been when it left the photon gun, making the light, again, indistinguishable from the light traveling down the other path. The second pair of sunglasses became known as the "quantum eraser" because once the method of distinguishing the light going down one path from the light going down the other path was removed, the wave pattern again developed on the screen. As in all other instances, the observer decided the outcome.

So what about all this? Does this mean that a wave that was traveling through space since before our parents were born just suddenly ceased to exist? Does it mean that we had a nice wave blithely traveling through the cosmos, minding its own business for almost a century until suddenly it was never really there?

Reality Check

> Yep.

See, that wasn't so bad. Today, many scientists around the globe are studying particle or wave "entanglement." There is no argument as to the reality of this phenomenon. Private companies are investing their own money into the research, looking for ways this technology can be of benefit to them.

As usual, the question is simple, and so is the answer. The problem is in *believing* it. That's another story entirely. Our neural pathways seem hardwired in the opposite direction. The same is true with the Bible. Many people also have a hard time believing what it says about the reality in which we live. There is, however, one thing about both science and the Bible that seems, to a large extent, to have gone unnoticed: the words are different; the concept is the same.

The scientific conclusion is that without an observer, there is no actual particle, just a wave. Heisenberg showed that once you do have an atomic particle—atom—it still doesn't mean that you are home free. His Uncertainty Principle (see chapter 15) states that the more you decide to know about the location of an atom, the less you can know about its momentum and vice versa. In addition to whether a "virtual" atom becomes an "actual" atom, to what extent an actual atom has an actual location or momentum is also entirely up to the observer. Without that observer, there exist only possibilities.

Hebrews 11:3 states, "We also know that what can be seen was made out of what cannot be seen" (CEV). Romans 4:17 says, "God … creates new things out of nothing" (NLT). In 2 Corinthians 5:17 we read, "For we walk by faith [what we believe], not by sight."

CHAPTER 45

IMAGINE

> A dream you dream together is reality.[101]
> —John Lennon

THERE IS A real problem with all this entanglement stuff, and I mean a *real* problem: the findings are irrefutable. The theory blows out of the water almost everything that has taken centuries for people to accept as scientific fact. Without an observer, there is no actual particle—no actual physical existence—but a wave. A wave is the means by which energy moves; it does not have any physical properties.

This means that we are living in an energy-based universe—energy that manifests itself as mass ($E=mc^2$). Before it becomes real, this energy exists as a "virtual" (possible) wave function or as a "virtual" particle. Which one depends on an observer. This light energy is defined as existing as a "probability wave," a "possible wave function," or a "virtual wave" unless or until an observer looks for possibilities. When that happens, the wave of energy "collapses" into what the observer is seeking.

If you are looking for particles, science says to imagine a "cloud" of millions of virtual (possible) particles. Each one represents a tendency or possibility that an actual particle might exist. Once someone measures or detects one of them, that single particle "pops" into actual existence and all the other possible particles instantly disappear.

Without a person—a source of information—the only things that exist are possibilities.

Reality Check

> Yes, it still is hard to come to grips with but, as we have seen, over eighty years worth of research hasn't been able to change the outcome. All it has done is conclusively confirm it.

If possibilities are all that exist without a person or information source, then one would think that the only limit on what can happen is the information coming from the source. Doesn't that mean that actually anything really is possible?[102]

"*And Jesus said to him ... 'All things are* possible *to him who believes'*" (Mark 9:23, emphasis added).

CHAPTER 46

SPACE AND TIME ARE NOT IMPASSABLE BARRIERS

> The stream of knowledge is heading towards a non-mechanical reality; the universe begins to look more like a great thought than like a machine. Mind no longer appears to be an accidental intruder into the realm of matter.... We ought rather hail it as the creator and governor of the realm of matter.
> —Sir James Hopwood Jeans[103]

WE SEE THAT it is true that atoms are not things; they are only tendencies (Werner Heisenberg) and also that faith (what one believes) "is the substance of things hoped for, the evidence of things not seen" (Heb. 11:1 KJV). According to the now-verified laws of physics, at the atomic level an observer causes reality. Quantum physics shows that, when it comes to atoms, an observer can affect not just the present or future reality but also the reality of the past. The problem for physicists, as well as the rest of us as we try to wrap our minds around all of this, is the realization that everything—big and small—is made out of atoms.

Let's go back to the experiment we set up in chapter 42. When we placed the mirror behind one of the holes in the first screen, the light going through that hole reflected in the opposite direction of the light going through the other hole. This meant that we had two waves traveling away from each other for ninety-nine years at the speed of light. Because the known classical laws of physics tell us that this is the fastest possible speed at which anything can travel, it means that if a particle detector was turned on ninety-nine years later, it would take at least 198 years for the detected particle to send some sort of signal to the wave going along the other path that it should disappear. Or if the detector did not detect anything, it means that it should take 198 years for some sort of signal to tell the wave that, because the detector was turned on but did not detect anything, a particle must appear on the screen at the end of that path.

The problem is that the data clearly show that even though a wave is hundreds of light years away, it will still disappear the moment a detector is turned on and a particle will appear on one screen or the other. Because the information about one is somehow immediately communicated to the other, the only logical assumption is that space and time are not the impassable barriers we thought they were.

The data prove that atoms are "entangled." They can and do affect each other in a way that ignores not only space but time as well. Each of us—the atoms that comprise us—are transcendent (non-local). In plain terms, this means that—theoretically—anything is possible. Reality is not just affected by the local surroundings but can also be immediately influenced by something that occurs on the other side of the galaxy.

At this point, it's time to stop and think about the person back in Galileo's time—let's call him Ed the farmer—who was having trouble taking it all in. There he was, watching the sun and moon pass overhead, as certain as anyone could possibly be that he was not moving. No other thought had ever entered his mind. Why should it? Everyone knows the difference between moving and being still. Ed and everyone he had ever known had fit all of life's experiences into this picture. At every moment of their lives, it was right before their eyes.

When Galileo announced that the earth was moving around the sun, the hard part for Ed was in believing it. It was easy to understand what was said; the hard part was walking out the door in the morning and accepting the perspective that he, the cows, and everybody else were rotating on a round ball and not falling off. Could he feel any movement? How could the wind and clouds be blowing east one day and west the next? Common sense told him—as it tells us about quantum physics—that the whole thing was utterly ridiculous. Copernicus almost gave up his work for the same reason.[104]

However, as time went by, the evidence could not be denied, especially when Magellan's crew set off one way around the world and came back the other. Quantum theory is crazy except for this same annoying detail: people keep coming back. They keep on successfully using it. The technological world we take for granted every day would be nonexistent without it. Our problem is not an inability to understand what is being said. Our problem, like Ed's, is trying to fit it into the long-held common sense notions of reality that we never thought would change.

For nearly four centuries Galileo's perspective of the universe has been expanded to where we have now come to see ourselves on a tiny speck, adrift in an unfathomable cosmos at the mercy of its dispassionate self-governing systems. Now that perspective has hit a roadblock. In order to understand the true nature of reality that science has conclusively proven, we have to once again find another point of view. Quantum physics now shows us that we are not insignificant creatures subject to the random forces of an immense universe after all. In fact, it says just the opposite. It says that the behavior of the basic component of all natural reality (atoms) depends on us.

Because the possibility of this truth has utterly and absolutely made no sense since it was first proposed in the 1920s, science, in the true spirit of the scientific method and skeptical inquiry, has done everything possible to disprove these findings. The fact that all the research conducted for nearly a century has done nothing but continually confirm this truth means that many scientists now realize it cannot be denied. As a result, as the twenty-first century moves forward, science is beginning to merge with what, since Galileo's time, has been seen as its polar opposite: the spiritual.

Every generation has looked at yesterday's truths as irrelevant nonsense. Why should we think that in the future people will think any differently about our current view of things?

CHAPTER 47

GOOD VIBRATIONS

> In essence, String Theory describes space and
> time, matter and energy, gravity and light,
> indeed all of God's creation ... as music.
> —Roy H. Williams[105]

AS WE HAVE previously discussed, in order to learn more about how we and all of life came into existence, scientists have attempted to understand the conditions that existed in the early universe just after the initial creation/Big Bang (whichever you prefer). One of the things this endeavor has brought to light is the amazing number of apparently unrelated coincidences that took place before anything we know of could exist. If any of a number of these fundamental conditions did not occur precisely as they did, there wouldn't be a universe. Here are a few examples:

- The strength of the positive energy charge in a proton is exactly equal and opposite the negative charge in an electron. Because there are billions of atoms in the smallest things we can see, the strength of this charge is so incredibly tiny that it really is impossible to imagine. However, if this delicate balance were off by just one in one hundred billion, not only would a grain

of sand explode, but so would we. No one has a clue how such incredibly small and precise equal and opposite values could just randomly happen.
- If the weight of an electron had been just slightly heavier or lighter, the nuclear process that formed stars (which formed the elements that formed us) never would have happened.
- Electrons are only allowed to orbit a nucleus at certain specific distances. This is like saying you could put a satellite in orbit around the earth at 1,000 or 5,000 miles, but nowhere in between. All distances from 1,001 through 4,999 miles are prohibited. No one has any idea how this strange condition came to be, but we do know that this phenomenon is what gives each element its unique characteristics, and that if this were not the case nothing could exist.
- Not only are electrons not permitted to orbit a nucleus except at certain specific distances, but they are also not allowed to *be* (exist) any distance from the nucleus except on those fixed avenues. To "change lanes" and get from one orbit to the other, they take a "quantum leap." As it turns out, when electrons change orbital paths, they simply disappear off one orbital path and reappear on the another without traveling through the space in between or taking any time to get there. No one knows how this could possibly happen. All we know is that if electrons were allowed to exist in between their allowable orbits—even if just passing through from one place to the next—atoms would either collapse or fly into pieces. We also know that this would not be good.
- The laws of physics, as we understand them today, tell us that the energy in the early universe should have produced an equal amount of matter and anti-matter. Anti-matter is interesting stuff. When anti-matter touches matter, a pinch produces an explosion larger than a nuclear weapon. This means that all of the matter and anti-matter produced by the Big Bang should have annihilated everything in existence a long time ago. However, for some unexplainable reason, for every billion particles of antimatter, the Big Bang produced one billion and one particles of matter. This turned out to be an extremely fortunate turn of

events, because that extra matter was just enough to produce you, me, and all the other stuff in the universe. Why or how this could have happened is anybody's guess.

- In his book, *A Brief History of Time,* Stephen Hawking observed, "If the rate of expansion at one second after the Big Bang had been smaller by one in one hundred thousand million, gravity would have caused the universe to have re-collapsed on itself. If it had had been faster by that same amount, the universe would have expanded into nothing by now."[106]
- A single neutron weighs about one tenth of one percent more than a proton. While that might not seem like much, it is crucial because this slight difference allows protons to remain stable. Neutrons decay because they weigh slightly more than protons, but because protons remain stable, it means that you, the table, and everything else will still be there when you wake up in the morning. What's puzzling is that because both neutrons and protons are made of the same things (quarks), there seems to be no discernable reason why one should weigh more than the other.
- In chapter 18 we noted that in this universe, like (meaning the same as each other) charges repel each other and opposites attract, except inside the nucleus of an atom. In there—and only in there—positively charged protons stick together. The force making this possible is called the "strong nuclear force." Although this force is strong when compared to the other things going on inside an atom, the term is misleading. If the force inside each atom that holds the protons together were infinitesimally stronger or weaker, not a single atom could have formed in the entire history of the universe. We think of the two other basic forces (gravity and electromagnetism) as being more powerful because they operate outside of atomic nuclei, but their values, or strengths, must be as meticulously precise. If the amount of energy that powers each were any different than exactly what it is, these fundamental forces could not interact with each other as they do, and everything in the universe would be a much bigger mess than it already is.

The idea that all of these incredibly precise values, each of which is fundamentally essential for all existence, could have spontaneously come about as the result of a big bang defies even the most implausible interpretations of any classical scientific theory. The reason and logic we have used to create the modern world we inhabit says it is simply impossible for all these things to have been the result of random events. This, coupled with the fact that classical physics has been unable to find any logical way for gravity and the quantum world to co-exist, has resulted in the emergence of three general categories of scientific explanations for our existence.

The first is called *multiverse* or *membrane* theory. This theory states that there are many universes (maybe billions or an infinite number) that exist like membranes or bubbles in a bathtub (in perhaps numerous parallel dimensions). Our universe is just one "membrane" or bubble in an infinite sea of bubbles that just happened to meet this incredibly coincidental set of conditions. In other words, there is more to existence than we can or likely ever will be able to detect with our five senses. As far as that statement goes, the Bible, of course, says that is correct.

A second explanation is *string theory*. Although this theory has been around for almost thirty years, it remains just that: a theory. In previous chapters, we have discussed how quantum mechanics says that all matter is in the form of waves of energy—atoms are not things, but only tendencies. String theory postulates—an intellectual term for "guesses"—that on a scale far smaller than atoms, there are tiny strings that vibrate to produce these waves of energy. Thus, according to this theory, all reality at its core is not made up of quarks, photons, and electrons but of far smaller vibrating strings of energy. Just as different vibrations—frequencies—cause different sounds, it is the different vibrations—tunings or pitch—of the tiny strings that cause all the various aspects of reality: in one instance a solid, in another a liquid, and so on.

If string theory is correct, in addition to the four dimensions of which we are aware, there are at least seven additional dimensions that house these strings that exist all around us. It is just that they are so small we can't see them. For instance, if you stand on the street and look at a telephone wire down the block, it will look as if it only has two dimensions. You can see that it definitely has length and a small amount of width, but if you didn't know any better, from your vantage point the line appears to be flat. It is only when you get much closer that you can tell that the wire is actually round and has another dimension: some depth. Physicists hypothesize that these other dimensions exist all around us, but even with the most powerful microscopes, they are still way too small to see.

A third theory is called the "Anthropic Principle." This theory states that since we exist, it is obvious that the universe has to be the way it is.

Reality Check

Duh!

No, really, this is a seriously discussed theory. That fact means this is scientific jargon at its ultimate. What this says is that we realize we really don't have any idea at all as to how all this came to be. All that we can really be sure of is that we do, in fact, exist. Let's give the principle that we can see ourselves in the mirror an intellectual-sounding name so people—especially our employers—will think what we are doing is important.

In his book, *A Brief History of Time,* Stephen Hawking said, "The remarkable fact is that the values of these numbers seem to have been very finely adjusted—tuned—to make possible the development of life."[107]

Perhaps that is because they were.

CHAPTER 48

TWO BASIC PREMISES

> But I fear ... your minds should be corrupted from
> the simplicity that is in Christ.
> —2 Cor. 11:3 KJV

IF YOU HAVE ever wondered why people believe what they do about the nature of our existence, you will find at the root of a person's conclusion one of two basic premises. Either the person believes that a supernatural existence beyond the power of human understanding lies behind everything we see and hear, or that the workings of the natural physical universe are the extent of all reality. Whenever the question relates to the nature of our reality and existence, the basis of a person's answer lies either in one place or the other.

A recent, widely publicized experiment illustrates this idea. In it, scientists directed some weak and otherwise harmless magnetic fields at a certain portion of a subject's brain. The woman was told to put on a wired helmet and relax for a little while. After the scientists directed these fields at her, she was asked what, if anything, she had experienced. The woman reported that there had been supernatural beings in the room with her.

In this case, the scientists filtered the data through one of the two presuppositions and concluded that this provided proof that the evolution of the brain had formed a specific area in order to compensate for the fear and anxiety produced by early man's first realizations of his mortality.[108] Of course, a person who believes in the existence of the supernatural would just as readily conclude that the effect of the magnetic field caused the person to sense what was actually there all along but typically goes unnoticed by the five senses. In either case, the person's interpretation of events rests on either one or the other of these two belief systems. As you read or watch people interpret scientific data you will see them—almost without thinking—immediately put their findings into either one or the other of these "world views." As quantum physics says, the person's reality will be caused by that person's own subjective view of existence. The scientist will have found what he or she was looking for.

Ever since Mr. Darwin shared his theory of evolution in the 1850s, these two basic ideas about the source of everything have clashed—sometimes bitterly and sometimes violently. Yet even though quantum physics has been a known aspect of reality since the 1920s, it has taken this long for some scientists—and fewer still in religious circles—to begin to see that the tension between these two points of view does not correctly frame the question. The plain and overwhelming evidence produced by quantum physics clearly states what, quite simply, no one has been willing to admit. The basis of our reality lies not outside us in the physical universe but *inside* each of us.

As we have examined in these pages, these ideas represent a drastic change in the way we have always been taught to see ourselves and our universe. They revolutionize our fundamental understanding of reality, causing us to throw out ideas that, for generations, no one had ever thought to question. We also have discovered the very real possibility that many things in the Bible that have remained shrouded in mystery for centuries may contain a lot more truth than anyone has, until now, been able to realize. As Jesus said in Mark 7:13, "You are nullifying and making void and of no effect [the authority of] the Word of God through your *tradition*, which you [in turn] hand on. And many things of this kind you are doing" (AMP).

CHAPTER 49

OUR CREATIVE POWER

> I believe that we get whatever we project.
> —John Lennon[109]

EVERYTHING THAT WAS created by humans on this planet was first imagined by someone who believed that the idea could become a part of this physical reality. First we have to imagine it—form a picture of the idea in our minds. Only then does the possibility exist that it will become a real physical thing. The power of imagination is limitless. As we saw in chapter 3, God pointed this out when He put a halt to the construction project at the Tower of Babel.

On the surface, if we put this idea about the power of imagination as it relates to God into the two basic premises discussed in the previous chapter, we will get one of two answers. God either created people with imaginations or He is a creation of people's imaginations.

But which is it?

Take a look around you. From the very first day someone opened his eyes and wondered what was happening all around, the one consistent thing that can be said about man's quest to understand his environment is that he has found it to be much more complex and astounding than anyone in his time ever had the ability to imagine, let alone comprehend. The latest discoveries do nothing but continue to confirm that. Throughout history, all people have ever really done is try to "smash" (tear apart) the pre-existing reality they have found themselves inhabiting in an effort to understand how it works.

And what have we found?

Everything we have ever thought we had finally figured out, turned out to be what the next generation called … wrong.

But now, as this new century unfolds, we are finally coming to grips with what Werner Heisenberg first noticed back in the 1920s: "Atoms are not things, they are only tendencies" (Scientific American [January 2005], 292, 34 doi:10.1038/scientificamerican0105-34). Not long after, Niels Bohr put it this way: "Everything we call real is made of things that cannot be regarded as real" ("Niels Bohr." BrainyQuote.com. Xplore Inc, 2012. 9 May. 2012. http://www.brainyquote.com/quotes/authors/n/niels_bohr.html). The concentrated research of the past eighty years has confirmed that the basic components of the physical reality we inhabit (atoms) require instructions before they will form the electromagnetic fields that comprise our physical universe. This statement is universal. It refers to all atoms. Not only does everything made by human beings require a conscious observer to first imagine and then make "real" what "are only tendencies," but the existence of every atom in the universe also adheres to the same requirement. This includes the atoms that make up you, me, and the galaxies on the other side of the universe. God … calls into existence the things that do not exist (Rom. 4:17 ESV).

Hebrews 11:1 states, "Now faith [what you believe] is the substance of things hoped for, the evidence of things not seen" (KJV). Proverbs 23:7 says, "For as he [man] thinks within himself, so he is."

Muhammad Ali, who rose from a poor upbringing in segregated Louisville, Kentucky, to become, arguably, the greatest heavyweight boxer of all time as well as an important and prominent force for social change in twentieth-century America, put it this way: "It's the repetition of affirmations that leads to belief. And once that belief becomes a deep conviction, things begin to happen" ("Muhammad Ali." BrainyQuote.com. Xplore Inc, 2012. 9 May. 2012. http://www.brainyquote.com/quotes/authors/m/muhammad_ali_2.html).

What we believe is the material of which our future is made. We have tremendous creative power. We get what we project.

CHAPTER 50

SEEK AND YOU WILL FIND

> Science is not only compatible with spirituality;
> it is a profound source of spirituality.
> —Carl Sagan

LET'S EXAMINE FROM a different angle the concept that what we believe creates our reality. Suppose you are hungry and there is a lake nearby. If you don't believe there are any fish in the lake, the last thing you are going to do is waste your time and money buying bait and getting your gear together to go stand on the shore to prove to yourself and everyone else that you are an idiot. In fact, if you don't believe there is any food in the grocery store, it will be the last place you are going to go when you're hungry. In the same way, if a person does not believe God exists, when that person has a need, he or she will view it as a waste of time seeking help from a myth. The point, as we discussed in chapter 47, is that you will not find it if you are not looking for it. In order to find out if anything is really there, you have to consider the possibility that what you are looking for can be found and then be willing to act—buy some bait, throw in a line, turn on a particle detector—in order to find out.

Here is another example of how what you believe creates your reality.

I have been a baseball fan ever since I was a little kid, and for the past twenty years or so my family and I have been attending our local team's games. Anyone who has ever been to a major-league baseball game knows that one of this endeavor's greatest obstacles is the parking. Where are we going to park? How much is it going to cost? How close to the stadium can we get?

One day the weather was nice, so we decided we would save some money and park a few blocks from the stadium. On our way, we drove right by the main entrance to the ballpark, and in the window of the little hut by the gate was a sign that said "free parking."

We drove on by, parked where we had decided to park, paid the attendant, walked down to the gate and, lo and behold, as a new promotion to the fans, the club was actually allowing free parking in the stadium parking lot right next to the ballpark.

What happened, you ask? What happened was that I have been going to these games for twenty years. I know all the parking lots like the back of my hand. I know how much extortion money I have handed over to those wily parking lot attendants. You didn't think for one minute, did you, that I was going to fall for someone's idea of a bad joke? Not a chance. I'm an old hand at this. I've been around the block (literally) more than a few times. Give me a break. Even if it wasn't a joke, the team had just returned from a long road trip, so someone—probably intentionally—had forgotten to take the sign down after some event that had occurred a few days ago.

Now suppose I had just arrived from some rural village in a third world country. Suppose I'd only been here for a few days and had absolutely no idea where I was or how I was going to get to this game. What do you think I would've done when I saw the free parking sign? That's exactly right: I would've pulled right in and thanked God for my good fortune.

The exact same thing happened when Jesus arrived on this planet. The religious people, who had spent *at least* a good twenty years studying the situation, already knew it all. They—like I was—were certain they already had the whole thing figured out. Convinced that they had seen all there was to see and knew so much more than some poor soul who could barely even read the language, they didn't have the ability to believe what was right in front of their eyes. However, those who didn't know a thing—beggars, prostitutes, etc.—had no problem whatsoever accepting and receiving the healing and blessing Jesus had for anyone who just pulled up with an open mind to see what was going on.[110] As far as the existence of God is concerned, Jesus said it is that simple.[111] When you see a sign, be open to the possibility that what it says could actually be true. "Seek, and you will find; knock, and it will be opened to you. For everyone who asks receives, and he who seeks finds, and to him who knocks it will be opened" (Matt. 7:78).[112]

CHAPTER 51

THE INFORMATION OF LOVE

> I think, therefore I am.
>
> —Rousseau[113]

Reality Check

You figured that out all by yourself?

"I THINK, THEREFORE I am" is the kind of statement that comes from philosophers and theoretical physicists who spend a lot of time sitting around pondering questions that can never be answered.[114] Who can do any more than make a wild guess at questions such as, "Why is there something instead of nothing?" or "Where did it all come from?" As Charles Darwin said (see chapter 18), "The mystery of the beginning of all things is insoluble by us."

So if we know we can never be certain about how all this came to be, the next thing for the scientist or philosopher to do is go back to square one and ask, "Well, what, actually, do we know? Is there anything I have no doubt about at all? Where is square one? What can I be absolutely certain is true?" Answer: "I'm sitting here thinking so I know I exist. I'm like pretty much 100 percent sure of that."

Reality Check

> "You cannot tell me that people actually get paid to think of that!" I told you back in chapter 39 to apply for federal grant money.

So, just in case there might possibly be someone on the planet who isn't 100 percent sure he is real, thanks to people like Rousseau,[115] he or she can now rest easy.

So where does that leave us? Right here feeling all sorts of things all the time. Each moment on this planet is part of a flow of sensation, both physical and emotional. We are immersed in a constant flow of energy. Our lives are about what we feel: hot, cold, love, hate, anxiety, peace, alone, appreciated, etc. What this means is that we spend our time doing what we believe will make that energy flow in a positive direction for ourselves, our families, and others.

One thing about each physical and emotional sensation is that it takes an opposite to define it. Making these distinctions is how we process everything, including who and what we are. If everything were just one temperature, there would be no such thing as hot or cold, because there would be nothing to compare with whatever we were touching. It takes one to define the other. Because of darkness, we understand light. It takes a negative to define what is positive. The more we know about hate, the more we understand what love really is.

We make these distinctions by decoding information as it flows to and through us. The smallest piece of information we know of is called a "bit." Depending on the source, a bit indicates either one of two possible things, each of which defines the other. A bit is either positive or negative, on or off, zero or one. Strung together, these bits of information flow via light—electromagnetic energy—throughout the universe creating and defining all existence, whether it is inside an atom, person, or a galaxy, or between them.

What we have seen in these pages is that it is the continuing flow of information via light—electromagnetic energy—that propels the universe. Science now says that this flow started with a bang about 13.7 billion years ago. The Bible says that flow was started by pre-existing information (see chapter 18). Although science does not use the term "supernatural" to describe how this energy flow began, what it does say is that the natural laws of physics did not exist before the Big Bang so there is no point in trying to use them to figure out whatever occurred before that.

The origin of the universe, then, is "beyond" the natural or is "super"-natural. This is a very recent scientific discovery. Less than one hundred years ago, the accepted truth was that the universe had always existed. Even our friend Al was at least fairly confident of that. We quoted him earlier as saying, "Only two things are eternal, the universe and human stupidity but I'm not sure about the former."

Because we can never scientifically prove how the universe got here, but we certainly do know we're sitting right in the middle of it, instead of "how," maybe it would be better to ask *what* we should do with this energy.

The universe gives us two choices: positive or negative.

Everything we do has a consequence. The problem is—and has always been—understanding exactly what has a positive and what has a negative consequence. Ideally, what we do as human beings is spend our time attempting to increase the energy flow toward the positive and then watch it benefit ourselves, our families, and the world. Yet as hard as it is to wrap our minds around it, we have seen people hijack airplanes and deliberately kill themselves along with hundreds of innocent men, women, and children, somehow convinced that this was a positive—good—thing to do. Somehow, young men and women are convinced that strapping bombs to their bodies, walking into crowded places, and blowing away themselves and many other unknown innocent people is good. No one ever acts upon a choice—investing his or her own energy, time, money, and especially life—unless he is convinced the result will be something positive for himself or someone else.

When God confronted Cain after he killed his brother, Abel, Cain's question was, "Am I my brother's keeper?" (Gen. 4:9). Another way to put it is, "What are you trying to say? You mean this is a bad—negative—thing?" The answer of course is *yes*! From there, the entire Bible goes on to tell the story of God attempting to put information as to what is good—positive—and what isn't[116] into the energy that flows between human beings.[117]

The fifth chapter of Galatians says that the only thing that matters is "faith working through love" (v. 6).[118] That is how positive energy enters the universe. Beginning inside us, we believe in love and, because we do, we consider it is worth the effort to expend our energy, speaking and acting to make it a reality.

Jesus did this to the ultimate. He was the embodiment of "good." He expended every ounce of energy He had to demonstrate that love is the most powerful form of energy in the universe. It is even more powerful than death. The question ever since has been, "Can anyone really believe that?" Is it possible that could actually be true?

When we look for an answer to that question, we must first remember that just because Jesus Himself was the embodiment of "good"—positive, love—that does not mean that the human beings who have represented Him throughout the centuries have always been the same. One look at church history—the Crusades, Inquisition, positions on race, mistreatment of children, etc.—provides more than proof of that. It cannot be argued that—throughout history right through to this present day—the behavior of many people doing things in God's name has chased people away from the one Whose name they proclaim.

As they run, there is one simple question: Where is the love? From their point of view, the only thing that really matters (see Galatians 5:6 above) is missing. The fact that those who represented Jesus' Father—God—did not love was the thing during Christ's life on earth that most upset Him. In fact, as you read through the Gospels, it appears to be the *only* thing He was upset about.

In the end, however, He remained positive. At the culmination of all that was said about Him and done to Him, he said, "Father, forgive them; for they do not know what they are doing."[119] His is a story of love overcoming everything, even death. He overcame the ultimate negative—an innocent agonizing death—with the ultimate positive: with love and forgiveness even in the midst of *all* the incredible pain and suffering as it was happening. It is this positive—this ultimate positive—that He invites us into.

"For God so loved the world, that He gave His only begotten Son, that whoever believes in Him shall not perish, but have eternal life" (John 3:16).

"These things I have spoken to you so that My *joy* may be in you, and that your joy may be made full" (John 15:11, emphasis added).

"Ask and you will receive, so that your *joy* may be made full" (John 16:24, emphasis added).

"And these things I speak in the world so that they may have My joy made full in themselves" (John 17:13).

"In him it is always Yes" (2 Cor. 1:19 ESV).

Remember that for reasons no one can scientifically explain, there is slightly more positive energy in the universe than negative, and, were this not the case, nothing, including life, could or would exist (see chapter 47).

CHAPTER 52

THE GREATEST TRUTH

> Love is a fruit in season at all times,
> and within reach of every hand.
> —Mother Teresa

IF YOU STILL find it hard to grasp what relativity and quantum physics have to say about the true nature of our universe, I can't say I blame you. It is my hope that in these pages you have found something of value—that something I have said has contributed to the positive as you travel your unique path.

We've noted numerous occasions in which the Bible defines God as light. It also uses another word to define God. The apostle John wrote, "He who does not love has not become acquainted with God [does not and never did know Him], for *God is love*" (1 John 4:8 AMP, emphasis added).[120] Here John is not saying that God is loving or that He loves us, which of course He does. Rather, he is saying that the very nature of God is love itself.

In the introduction to this book, I wrote about the skepticism I had earlier in my life toward Christianity. My well thought-out conclusion—at least as I saw it at the time—was that most Christians were delusional dreamers and that the born-again variety were borderline psychotic. However, when I read this simple statement—"God is love"—I thought, *You know something, I like love. If that's what God is, I'm all for it.* My reasoning was, *I know that light is real and love is real. If that is what God is, I believe that.* All hearts search for and desire love. If God is love, all hearts search for and desire Him.

In this book, we have attempted to peer through the maze of information, theory, and speculation at what have been, until now, the secrets of the universe. There is, however, a bigger secret that somehow has remained hidden behind that same maze of information, theory, and speculation, not to mention the lust men have for power, attention, money, and all that comes with them. The biggest secret—and, therefore, the greatest revelation—is the fact that *God is love*. The greatest truth is that real unconditional love exists (see John 3:16-17)[121] and that it can be a reality in your life or anyone's life.

The list of reasons this concept has remained hidden from most of the world is as long as the two thousand years since Jesus first demonstrated its reality. In 2 Corinthians 11:3, the apostle Paul lamented, "But I fear, lest by any means ... your minds should be corrupted from the simplicity that is in Christ" (KJV), because at its heart, the nature of this universe is really very simple. Like the search for any truth that we have discussed in this book, all you have to do is be open to the possibility that what you are looking for does exist: a real love that comforts you in a way that only God, as your Father, can. Then take a step in that direction and see what happens.[122] Note that a man named Jesus went to an awful lot of trouble to prove that His Father is real and that His Father is also your Father. As a real child, you can rest in His acceptance of you as His own simply because of who you are. And because you really are His own child, you have great value and purpose.

As Mother Teresa said, Jesus really did die to put "love within the reach of every hand." He waits to delight in you.

"May the Lord bless you, and keep you; The Lord make His face shine on you, And be gracious to you; The Lord lift up His countenance on you, And give you peace" (Num. 6:24-26).

ENDNOTES

1. John Updike, *Roger's Version* (New York: Alfred R. Knopf, 1989), p. 10.
2. Niels Bohr (1885–1962) was a Danish physicist and founding father of quantum physics. By the twentieth century, no single person could possibly develop an entire branch of scientific research on his or her own. However, with the exception of Albert Einstein, Niels Bohr probably contributed more to our currrent understanding of the universe than anyone else during the past one hundred years.
3. There is the exception of that certain segment of the population who, as a service to the rest of us, already has everything figured out. This includes a small percentage of intellectuals, scientists, lawyers, journalists, theologians, and virtually everyone aged thirteen to nineteen.
4. Like with a mop.
5. Voyager 1.
6. 1934–1996. Astronomer, astrophysicist, and author, whose greatest contribution may have been the late 1970s PBS series *Cosmos,* which did so much to popularize and give a fundamental understanding of modern science and cosmology to the general public not only in the United States, but also throughout the world.
7. You can buy a similar one today for about $30.

8. Stephen Jay Gould (1941-2002) was a paleontologist, evolutionary biologist, science historian, Harvard professor, author, and one of the most influential evolutionary spokesmen of his time.
9. Do not kill, do not steal, and so forth.
10. If it seems presumptuous to call Albert Einstein "our friend," we will see (in the chapters to come) that it really is a fitting description. The things he contributed to mankind are rivaled by very few. Today, these scientific revelations are so much a part of our daily lives that we (myself included) simply take them for granted.
11. 1859–1941. French philosopher.
12. He thought he was in India, when in fact he was actually on the opposite side of the planet. If you find it hard to believe that anyone that famous could be so misplaced, Columbus's lasting legacy proves it. In the U.S., we have a state, a major league baseball team, hundreds of high school and little league teams, as well as numerous towns and cities—not to mention an entire race of people (in recent years more aptly called Native Americans)—all named after the inhabitants of a country who were living on the other side of the planet. Personally, I take great comfort in this and find Columbus Day to be one of my favorite holidays. I can't count how many times the thought of ol' Chris comforted me as I said to my family, "Okay, we're lost, but I have yet to blow it as to what hemisphere we're in." Although I poke fun, Columbus did "boldly go where no one from his world had gone before seeking new life and civilizations."
13. Personally, I don't buy the story that Magellan's argument with a local tribal chief caused his death. After all, who had he been talking to before he left? When Mr.—who, me, ask for directions?—Columbus got home, what did he tell everybody? He told them he had been to India. So for Magellan's crew, the Pacific Ocean came as *just a bit* of a surprise. I definitely believe that it was his crew saying, "When are we going to get there?" that did him in.
14. A fictional character played by Art Carney in the TV series *The Honeymooners*, who, during a golf lesson, was told to address the ball.
15. Okay, more dizzy than they already are.

16. Not advisable, according to Proverbs 3:5.
17. Hal Lindsay, *The Road to Holocaust* (New York: Bantam, 1990), p. 7.
18. Luther managed to escape with his life. Others were not as fortunate.
19. This same phenomenon—public access to new information—was again responsible for massive social change toward the end of the twentieth century. After the Second World War, Communist countries set up what became known as the "Iron Curtain." Its purpose was to keep people in and information out in order to maintain control of the populace. However, by the end of the 1980s the spread of ideas through new information highways—satellite, TV, radio, video tape, etc.—became so pervasive that the people—as happened when folks were able to read the Bible for themselves—had access to enough information to enable them to believe they were fully capable of deciding the truth of matters for themselves. As of this writing, North Korea, the last bastion of what some have now called the "Bamboo Curtain," is in the process of dying a slow death. Also, at this time—the start of the twenty-first century's second decade—this same process—free public access to information, now via the Internet—is having a profound transforming effect in the Middle East, China, Africa, and other cultures where access to information had been restricted by government or was otherwise unavailable.
20. Unfortunately, at that time, the definition of the term "men" did not include "wo-men." And when those with a darker skin tone attempted to appropriate this truth for themselves, they were declared—by the nineteenth-century U.S. Supreme Court's Dredd Scott decision—to be not fully human, which made it perfectly legal to treat them as you would any animal you owned. So although much was lacking in 1776 and we certainly have yet to arrive, if there is one thing we can say, it is that the populace of the entire world is now more aware than it has ever been of the truth Jesus died to proclaim. Each and every person has great value and certain unalienable rights because they are created by, and therefore equally loved by, God. Because everyone is a child of the same Father (God),

there is not one who can say he is better than another, and each has an obligation to treat the other as he would himself.
21. We sense length, width, and depth—but not time—with our eyes. Time is more a perception of beats: rhythm. As we go on we will see that rhythm—in terms of frequency, etc.—really is a fundamental component of reality.
22. This is all perfectly correct, logical, and very sound advice. However, there is a greater, more fundamental, and important truth at the heart of Jesus' purpose on earth. That is His personal relationship with you. While He was here, not surprisingly, Jesus was asked a lot of questions. One of them was, "What is the most important commandment"—the most important thing we should do? *THE MESSAGE* translation puts His answer this way: "Love the Lord your God with all your passion and prayer and intelligence" (Matt. 22:37). The *Amplified Version* is probably more familiar: "You shall love the Lord your God with all your heart and with all your soul and with all your mind (intellect)." Jesus then said, "This is the great (most important, principal) and first commandment" (Matt. 22:38, AMP). This *most* important principle clearly requires taking your attention off of the things of this world, which is not something humans seem to naturally think they should be doing. Even though it is the most important commandment, unlike the second ("love your neighbor as [you do] yourself"; Matt. 28:39), it is rarely, if ever, talked about. Yet it states the primary reason for Jesus' life and death. Although what you do—the second commandment—is important, the *most important* thing to God is His personal relationship—interaction—with you. The most important thing is that He loves you and that you experience your own unique individual relationship with Him. Jesus sacrificed all to make this possible. He died so that He could establish His Father (God) in your heart as your Father (see John 17:13-21). He is saying that, just as it is with your own children, it is not what you do but who you are—a child of a Father who desires, more than anything, to spend time with His son or daughter—that makes you so valuable to Him. As a true loving Father, no matter what you have done, He loves you, forgives you, and is hoping, waiting, for you to come home (see Luke 15:11-32).

When Jesus told His disciples that He was about to leave this earth, they implored Him not to go, believing that they would be abandoned and lost without Him. But He told them that His leaving was not a bad thing because it would establish a way for them to have a permanent relationship with His Father. He said, "I will ask the Father, and He will give you another Comforter (Counselor, Helper, Intercessor, Advocate, Strengthener, and Standby), *that He may remain with you forever*—the Spirit of Truth, Whom the world cannot receive (welcome, take to its heart), because it does not see Him or know and recognize Him. But you know and recognize Him, for *He lives with you [constantly] and will be **in** you. I will not leave you as orphans* [comfortless, forlorn, helpless]; I will come [back] to you" (John 14:16-18 AMP, emphasis added).

Jesus told His disciples that His Father was their Father too. He said that He would not leave them as children without a Father—orphans—but would provide a means by which they would have their own personal permanent relationship with their Father God. This is the mystery Paul spoke of in Colossians 1:27 when he said, "God willed to make known what is the riches of the glory of this mystery among the Gentiles, which is *Christ in you,* the hope of glory" (emphasis added). Although this sounds like a mystical concept, it really is not. If you have a family, you have a bond unlike any relationship you have with anyone else on the planet. Jesus went through all He did to make it possible for each of us to establish this bond—the same one He had—with our Father in heaven.

If you don't have this relationship in your life and you want it, then just as with any loving Father, go to Him and ask for it. I think a lot of people are like I was. They make up their mind that it's all a bunch of nonsense before ever trying it out. It doesn't take much of a risk to seek out answers unless, as in my case, pride—the fact that I believed I already knew it all—gets in the way. Many others feel that because no one has ever cared for them before, no one would want to care for them now. After feeling hurt and rejected most of their lives, they are—understandably—afraid (almost like a reflex action) to open their hearts for fear they will just get rejected again. But that's why Jesus suffered such rejection Himself. So that this time, when we ask, the answer will actually be, "Yes, I really do love you."

23. As first noted in chapter 9, this is where the idea that all people have equal and unalienable rights originated. By definition, evolution is about inequality. The processes of nature select the fittest—those who are better than others. However, as God's children we all—weak and strong—are a human family, and just as all children love and take care of their siblings, this status requires all of us to treat others as we would want to be treated. As we saw in chapter 9, no one believed that a few hundred (let alone two thousand) years ago. A person's worth was based on ancestry, nationality, sex, and other factors. The idea that the Son of God voluntarily suffered for the wrongs of others in order to restore each person (child) to a loving relationship with his or her Father (God) was and still is unique to all religious thought. Unlike any other spiritual or metaphysical view, it places extraordinary value on each individual—an individual who is equally loved and valued, not because of what he or she may do or not have done, but because of who he or she is (a precious child of the Father/God of the universe).

Old ingrained perspectives, however, do not disappear overnight. Two thousand years ago, few were able to grasp the idea. It was barely self-evident in 1776, years later. Many more years had to pass before women and people of different ethnicities would be legally included in the concept. Yet the continuing proliferation of this conviction enabled the United States to elect a president of African-American heritage and continues to be a model for personal rights, worth, and the intrinsic value of each human being around the world. Attempts to impose democracy in some corners of the planet have been dismal failures due to the lack of this fundamental belief in a majority of the people living there. That should not be hard for government leaders to understand. Our own history shows that the success of this first attempt to form such a society here in America was anything but a slam-dunk, as Abraham Lincoln pointed out so well in his Gettysburg Address.

24. Numbers like that are hard to fathom. If a person lives to be seventy-five years old, that's only about 39.5 million minutes. That means seventy-five years is less than 2.4 billion seconds, not two hundred billion. It takes more than six thousand years for two hundred billion seconds to tick by. It really is hard to picture such a number. At this writing, there are close to seven billion people on the earth. Including us, estimates are that approximately one hundred billion people have been born and lived on this planet since humans first appeared on it. That is about two galaxies (each with billions of stars) for each person who has ever existed.
25. And you thought you didn't possess a scientific mind.
26. Richard Feynman (1918–1988) was a Nobel Prize winner who is considered by many to be one of the greatest physicists of the twentieth century. His lectures, which are more popular now than they were during his lifetime, are widely available in both audio and written form. One of his most popular and aptly titled books is *Surely, You're Joking, Mr. Feynman!*
27. This in itself was a novelty, as it marked the first time a scientist publicly admitted being uncertain of anything since Ben Franklin said he wasn't really sure what would happen when he launched a kite in a thunderstorm.
28. This is the speed that light travels. It is generally accepted that nothing can travel any faster, although the entanglement of atomic particles that we will discuss later suggests that may not be true.
29. Stephen W. Hawking, *The Theory of Everything* (Beverly Hills, Calif.: New Millennium Press, 2003), p. 36.
30. The Bible is silent as to the acoustics of the event. The "noise" of the Big Bang is thought to be heard today in what is called the "Cosmic Background Radiation." When a TV is turned on and no broadcast signal is coming through the cable, it is not silent. Some of the static seen and noise heard is thought to be part of the Cosmic Background Radiation, which many believe is the echo of the Big Bang.
31. It is interesting to note that after this initial creation of light, Genesis 1:5 states, "there was evening and there was morning, one day." Whatever evening, morning, and day it was, it was not measured by

the sun or moon. Because a "day" is something that measures time, it is possible that this creation of "the first day" (Gen. 1:5 KJV) could be a reference to the creation of time (see Heb. 1:2). Incidentally, Plato (427–347 B.C.) wrote, "Time, accordingly, was created along with the heavens; in order that, coming into being together, they might also be together dissolved, if ever their dissolution should take place." (*The Ante-Nicene Fathers*, vol. 1, PC Study Bible formatted electronic database, copyright © 2003, 2006 by Biblesoft, Inc.)

32. Stephen W. Hawking, *The Theory of Everything* (Beverly Hills, Calif.: New Millennium Press, 2003), p. 42.

33. Since this discovery during the first half of the twentieth century, human beings have realized that duplicating this process would solve the world's energy problems. However, to this date, scientists have been unable to intentionally replicate this apparently unintentional occurrence, even on a small scale.

34. Although astronomers state that, on average, a galaxy produces a supernova once every fifty years or so, modern astronomers have never witnessed one in our own Milky Way Galaxy. The last was observed by Johannes Kepler in 1604. Although not directly observed, residual evidence has been detected of two other possible supernovas in our galaxy in 1680 and 1868.

35. Francis Crick, Nobel Prize winner for discovering DNA (one of the most valuable and noteworthy discoveries in the history of science), concluded that its function was so complex that it could never have occurred by chance. DNA contains the instructions that built the proteins that built the molecules that built you. Imagine a single cell as a manufacturing plant that receives a myriad of orders and produces a thousand different products in a single second. Each product (protein) consists of hundreds of parts (amino acids) that are precisely assembled, checked for errors, and is either used inside the plant to build other products or shipped to numerous customers. From an evolutionary perspective, the function of a single cell (fundamental component of all life) is puzzling. One problem with this assembly process is that the plant cannot run without the products it makes. Cells use proteins, but cells also make proteins. No one appears to know how one could first exist without the other.

Another problem is that half a cell is as functional as half a car. Evolution says that primitive cells gradually became more complex. To do this, a cell would have needed to be able to function in a rudimentary state. But we now know this is not possible. Like a car, it cannot move with half an engine.

Each cell has millions of employees, none of whom ever take any breaks. They are packed so tight that L.A. traffic might as well be Colorado during the snow. How do they avoid gridlock? Cellular engines are pretty high tech. A simple single bacteria cell runs on nuclear fuel (proton power) at anywhere from seventeen thousand to one hundred thousand RPMs. But it's the fact that a cell assembles and repairs itself that makes me wonder why I have to go to the used (sorry, "pre-owned") car lot. These engines (cells) are so small you have to line up about twenty-five thousand of them to make an inch. Still, that only addresses the hardware. The real question is where did the information (software—program instructions) in the DNA come from? How do all the employees know where to go and what to do? How does the cell know how to take the atoms of carbon, hydrogen, oxygen, etc., and create the proteins? Who writes the invoices directing these finished products (proteins) to their precise destinations? How does a cell know how to then reproduce itself?

Then again, how do the sixty-four cells that constitute a fourteen-day-old human embryo know how to produce you? There are so many instructions in human DNA that if the DNA in your body were uncoiled into a straight line, it would stretch from here to the planet edge of our solar system and back (about ten billion miles). Seeing all this, Mr. Crick proposed an answer. He called it "Directed Transpermia." He concluded that since it unquestionably is impossible for such a thing to occur spontaneously, aliens must have seeded this planet. Due to his stature as one of the most important scientists of the century, he was taken seriously. From a biblical perspective he should have been, as the Bible says a non-resident alien (God) did exactly that.

36. In his book, *The Universe in a Nutshell*, Stephen Hawking compared the information in today's most advanced computers

to the complexity of an earthworm's brain. The information that comprises just one of us would fill enough terabyte (one thousand-gigabyte) hard drives to fill up five hundred million Empire State buildings. No wonder we need to spend one-third of our lives defragmenting (sleeping, resting, and reorganizing).

37. Realizing that nothing can happen in the universe without information, a Massachusetts Institute of Technology professor, Seth Lloyd, wrote a book entitled *Programming the Universe*. In it he postulates that the universe is actually a quantum computer. However, where the information that the universe computes came from remains unknown. He goes on to say that it is, therefore, theoretically possible that someday we could build a computer that could simulate the universe to the point that the simulated universe would be indistinguishable from the real. To put it another way, he is saying that someone's mind could create a universe just like the one we live in. This, of course, is in agreement with the Bible, which says that the mind of God did exactly that.

38. Black holes were first called dark stars. If you throw a baseball up into the air, it will only go so far before gravity pulls it back down to earth. To continue going up and escape the earth's gravity, the ball would have to be thrown about twenty-five thousand miles per hour—the speed a rocket travels to launch people and things into space. Light travels a lot faster than twenty-five thousand miles per hour (186,282 miles per second), so it has no trouble leaving the earth. You could fit almost a million earths into the sun, but light still has no trouble leaving it. Imagine, however, an object in space so dense that its gravitational pull will not even let light leave its surface. Just like a baseball on earth, the light photons get pulled back to the surface before they can get very far. Hence, from our point of view, we see a black hole. Although black holes cannot be seen, their immense gravitational effects on objects near them certainly can. The Hubble space telescope has confirmed the existence of black holes. They appear to exist at the center of most major galaxies, including our own Milky Way. As to why they are there, no one knows.

39. Frequency is a measure of how often the crest of one wave, and then the crest of the next wave, passes a certain point. This naturally depends on the shape and size of a wave. The larger or longer the shape of the wave, the lower the frequency. The higher the frequency, the shorter the length of the wave. The crest of a wave passing a certain point is also a measure of time. The flow of electromagnetism is also the flow of time.
40. Until recently, these fundamental forces had been divided into four categories. The fourth is called the "weak nuclear force." This force only works inside the nucleus of an atom and is responsible for radioactive decay. Recent research, however, has shown that the weak nuclear force is actually an aspect of the electromagnetic force.
41. This produces energy and is how handheld calculators, solar panels, and similar instruments are powered today.
42. A German physicist, Max Planck, had earlier found that energy seemed to travel in small packets (or quanta) of energy. Einstein concluded that he was looking at what Planck had described on the photoelectric plates.
43. Yes, this was before the advent of primetime television.
44. When an electron absorbs or emits a photon, the amount of energy in the electron either increases or decreases. This causes the electron to alter its orbit. It either changes orbits around the nucleus where it is, or it leaps to an orbit in another atom. It also means the energy level of the atom receiving a photon is increased and the energy level of the atom losing the photon is decreased. This means the physical state of the atom has changed and, therefore, so has the physical state of whatever the atoms comprise. Atoms seek to retain their naturally balanced light—electrically charged—condition. For example, if there is a sodium atom missing an electron and a chlorine atom with an extra electron, they will combine to balance themselves and instead of two toxic elements the result is sodium chloride: salt.
45. So do Bible believers.

46. Roger McGuinn and Chris Hillman, "Change Is Now," from *The Notorious Byrd Brothers*, 1968, audio CD, March 25, 1997, Columbia/Legacy CK 65151.
47. If you have never been to the Mall in Washington, D.C., I highly recommend it. There are many incredible things to see, the most astonishing of which is our tax dollars being well spent. The Air and Space Museum and Lincoln Memorial are just the beginning. And it's completely free—except, of course, for the parking.
48. My son, who was sitting next to me, didn't say a word. However, the next week after we got back, he announced it would be better if he got out of the car a block before we actually pulled into the high school parking lot.
49. Didn't you know that tiny particles called neutrinos are traveling through you, the entire earth, and out the other side of the planet without ever hitting a thing? That is, unless they happen to be going through New York, in which case they have to stop and pay a toll.
50. What you "hear" someone else "say" and relate to another person is called "hearsay" in the legal business. Its inherent unreliability is the reason (with a few exceptions) that it is not allowed as evidence in the courtroom.
51. There is a finite amount of water in the sea, ground, icecaps, plants, atmosphere, people (who are primarily composed of water), etc.—about one million cubic kilometers; the same amount as in the time of Job. It has been recycling ever since. As to where water came from, its origin on earth remains one of science's great unsolved mysteries. Interestingly, the Bible says that God separated *the waters* and called one part heaven and the other the seas of the earth (see Gen. 1:6-10).
52. You may have heard this referred to as the "space-time continuum."
53. He has no idea how old his kids are, but this he can't forget.
54. This knowledge caused him to invent the lightning rod. Although he certainly could have applied for a patent, he did not. In this case, instead of pennies, he saved property and lives.
55. Examples range from hydrogen, which consists of one proton and one electron, to uranium, which has 92 protons, 146 neutrons, and

92 electrons. Each atom contains the same amount of negatively charged electrons and positively charged protons with precisely balanced equal and opposite charges. Scientists have produced other elements by combining protons, neutrons, and electrons; however, unlike the 90 that occur naturally, these artificial elements are unstable and eventually fall apart.

56. Too bad the same thing doesn't happen with Democrats and Republicans. One hundred years ago, working conditions were deplorable and the Democrats were certainly right to reign in and regulate big business. Of course, Republicans are also correct when they say that overdoing that destroys the prosperity and jobs everyone needs. The problem is not policy. The problem shows up, whether it's lying on tax returns to collect welfare or to own ten cars instead of nine. Or if you are actually writing the policy, doing it in a way so that ten of the sixteen most prosperous counties in the United States—including the top three—all just happen to surround Washington, D.C. The more that people cheat, steal, threaten, and hurt each other, the less freedom any society has.
57. Roger S. Jones, *Physics For The Rest of Us* (Chicago, Ill.: Contemporary Books, 1992), p. 7.
58. At minimum, the number of stars in the Milky Way Galaxy.
59. Mark Twain (1835–1910) wrote fiction. His manuscripts had to make sense or they wouldn't sell. The truth is under no such constraint. History shows us that the truth of the matter has always been a hard sell—usually only grudgingly accepted after long arguments until decades, or even centuries, have left no viable alternative.
60. Richard Feynman (see chapter 15) likened this to taking two wristwatches and smashing them to pieces in order to figure out how they work.
61. I was perfectly happy with just one on *Star Trek Deep Space Nine*. He was charming and strange all by himself.
62. *The Harmony of the World* is the title of a book written by Johannes Kepler in 1618. It is said that he regarded it as his most important work.

63. Each note created by a guitar or piano string consists of its fundamental tone and resonating harmonics (waves that correspond with the fundamental tone), which give it its full sound. Without them, the note is flat and boring, like what you would hear from a plucked piece of string.
64. Hinduism and Buddhism attempt to describe the concept and the sound of this universal harmony with the words "Om" or "Aum."
65. See chapter 27.
66. During the 1990s, the United States began constructing a fifty-four-mile tunnel under Texas for the same purpose. Congress killed the project in 1993 but not, of course, until more than two billion dollars of someone's income (tax money) had already been spent digging a fourteen-mile tunnel to nowhere.
67. Something that theologians and scholars do not seem to have the ability to do, probably because they wouldn't have a job if they told everyone, "Hey, it just means exactly what it says."
68. Ignoring the odds, one adventurous soul by the name of Moses did attempt to approach God. Scripture states that when he came down Mount Sinai after receiving the Ten Commandments, "His face shone and sent forth beams by reason of his speaking with the Lord" (Exod. 34:29 AMP).
69. Pronounced the same as "ether" used in anesthesia. The term had its origin in Greek mythology referring to the special type of air in the upper sky or heavens that the gods breathed.
70. Big deal. Why shouldn't they know how it feels too?
71. Sound reasoning, don't you think? Al's colleagues, however, would never admit it. As noted in chapter 31, this was the revelation that unlocked relativity. But instead of giving him a Nobel Prize for one of the single greatest intellectual achievements in human history, the scientific establishment gave him one for his work with the photoelectric effect discussed in chapter 20. As noted in chapter 27, our friend Al didn't even discover the photoelectric effect.
72. Tattoo on the right arm of Derrick Rose, Point Guard for the Chicago Bulls, NBA, 2011.
73. Stephen W. Hawking, *The Theory of Everything* (Beverly Hills, Calif.: New Millennium Press, 2003), p. 54.

74. Yes, we have also described light moving as a wave. We will discuss this unique "dual" nature of light later.
75. Photon guns do exist as we will see in Chapter 39.
76. John McEnroe.
77. I told you there was still hope for the rest of us.
78. Douglas Adams (1952–2001) was a British comedian and writer.
79. You had always suspected this about certain people, and now, finally, here is the confirmation.
80. At the speeds we live our lives, such differences are incredibly small, but they do exist.
81. Astronomers routinely gaze at galaxies more than twice that distance from us.
82. Your actual speed depends on where you are on the earth. At the equator, you are moving at one thousand MPH. At the north and south poles, not very much.
83. That is faster than a trip from New York to Los Angeles to Washington, D.C., and then to San Francisco … in one minute.
84. See chapter 36.
85. The kind of thinking that, to this day, keeps the federal grant money rolling in.
86. This was the first photon torpedo. And you thought *Star Trek* was pure science fiction—or should I say, most of you. In 1865, Jules Verne wrote about three men who blasted off from Florida in an aluminum capsule on a four-day trip to the moon and then splashed down in the Pacific Ocean. This was pure science fiction until 1969, when it actually happened.
87. Earlier, we talked about how space and time are actually two different aspects of the same thing that scientists call the fabric of space or the space-time continuum (see chapters 22 and 38). Also, remember our discussion of electromagnetic energy in chapter 19? This energy is the force behind everything that happens in the universe, unless it is caused by gravity or operating inside the nucleus of an atom. Electromagnetic energy flows in waves. Another way to describe waves is by their frequency. Frequency is a measure of time. Because atoms have a wave-like nature, that means everything made of them, like you and me, has a frequency

or rhythm. Whatever is happening—your heart beating, atoms moving, galaxies rotating billions of light years from here—there is a rhythm to it. When trying to describe the space-time continuum in chapter 38, we said that you can't be in space without taking time to do it. Time is frequency; rhythm. We are creatures of rhythm living in a rhythmic, pulsating (electromagnetic) universe. What is it about music that affects us so? When we say that it resonates with us, maybe that is *exactly* what it is doing. Maybe thinking about rhythm is not only a good way to think about the unity of space and time, but also a good way to think about the unity of us all. Good music has a wonderful ability to bring together what, otherwise, would be the most peculiar group of people one could ever imagine.

88. Let's face it: some folks just seem to be possessed with an impediment to their thinking called "logic." My personal opinion is that these people have never been married, because anyone who has experienced this aspect of reality is well aware that in order for one of the most basic and necessary of nature's systems to function, logic must be abandoned every step of the way. Hold on to logic and what do you get? Poor, lonely, thoroughly confused people and rich lawyers. Abandon it in a marriage relationship and out of nowhere you suddenly find peace, harmony, growth, and positive flow. And what are people made of? *Atoms*, of course. I rest my case.

89. "In this big crowd with so many people pushing and pressing in on us, how could anyone possibly know who touched you?"

90. As you know, Jesus was also called "Christ." Christ is a Greek word that, when translated, means "Anointed." In those days people didn't have last names, so they were identified by what was unique about them: John the "*Baptist,*" James the "*son of Zebedee*" (Matt. 4:21). Magdalene was not Mary's last name. It was a reference to Magdala, the town she was from. Jesus "*Christ*" was the guy Who went around with anointing (power) healing people. As far as *why* this one word, and only this one word, was not translated from Greek into English, that I will have to leave to your fertile imagination.

91. "Who is that? Jesus? I knew that kid when he was growing up. Give me a break."
92. To illustrate the point in a different way, suppose you have a very close friend who (for whatever reason) you believe doesn't like you anymore. If you see him or her out with another person, you will interpret that as confirmation of your opinion. If you then express your unhappiness or anger a few times, even though it actually may have been completely untrue to begin with, your belief will start to create that reality because no one likes to be around a person who is angry with him or her for no apparent reason.
93. John Wheeler (1911–2008) was a leading twentieth-century physicist and a Princeton professor who participated in the first atomic bomb experiments and taught future physicists, such as Richard Feynman.
94. I realize that those outside the intellectual establishment assume this to be the norm, but just like membership in any other exclusive group, adherence and faithfulness to the group's long-held beliefs and traditions are a primary requirement for belonging. Thinking not in line with those traditions is, at best, highly suspicious and often leads to what is described as the "left foot of fellowship," which is a way of illustrating what is applied to your rear end as you are escorted to the door.
95. Others weren't nearly so kind.
96. See chapter 9.
97. Approximately 5,874,601,673,544 miles (a short distance by universal standards).
98. Just as fast as any college professor (or maybe even faster, since you don't have so much unlearning to do).
99. Actually, the technical description is that "the wave function collapses." But that's just the scientific term for "poof."
100. See V. Jacques, et al., "Experimental Realization of Wheeler's GedankenExperiment," *Science* 315:966 (2007), e-print at http://www.arxiv.org/abs/quant-ph/0610241; John Cartwright, *Physics World*, "Photons Denied a Glimpse at Their Observer" (February 15, 2007), http://physicsworld.com/cws/article/news/27106.

101. "John Lennon," BrainyQuote.com, Xplore Inc, October 25, 2011. http://www.brainyquote.com/quotes/authors/j/john_lennon.html.
102. Proved in 2004 when the Boston Red Sox won the World Series.
103. James Hopwood Jeans (1877–1946) was an English physicist, astronomer, and mathematician.
104. "When I considered this carefully, the contempt which I had to fear because of the novelty and apparent absurdity of my view, nearly induced me to abandon utterly the work I had begun." Dedication of the Revolutions of the Heavenly Bodies to Pope Paul III, Nicolaus Copernicus (1543)
105. Roy H. Williams is a businessman and bestselling author who has a popular website: www.mondaymorningmemo.com.
106. Stephen W. Hawking, *A Brief History of Time* (New York: Bantam, 1998), p. 121.
107. Ibid., p. 125.
108. This, indcidentally, is the same conclusion that has been reached by almost all secular scientists and philosophers who have been trying to understand the source of religion ever since Charles Darwin first published *The Origin of Species*.
109. John Lennon, quoted in the *Jim Ladd Interviews*, October 10, 1974.
110. Nothing has changed, by the way. Today, God is moving in new incredible and miraculous ways in Africa, South America, Asia, and many other places in the world where people are simply willing to come and gladly accept whatever God wants to do for them right there right then. But what about the traditional church in the countries where we already know, where we have already thoroughly studied the situation and are certain (like I was) we already have it all figured out? What do they say about all this? Do they even have eyes to see (see Matt. 13:15)? Or are their traditions preventing them from having any idea what is really going on (see Mark 7: 6-13)? Whether it's science, religion, or looking for a place to park, pride is a tough taskmaster. "*Thank you, Father, Lord of heaven and earth. You've concealed your ways from sophisticates and know-it-alls, but spelled them out clearly to ordinary people. Yes, Father, that's the way you like to work*" (Matt. 11:25 MSG).
111. Please also see chapter 42.

112. If you can't help but wonder why, if it's so simple, it often seems so complicated, then let me reiterate what I said in the previous chapter, which is that all too often there has been a huge difference between what people say about God and what He has said about Himself. Given the way people behave (and have behaved throughout history) in the name of God, it's no wonder that many run the other way. As I pointed out, this inconsistency is a primary theme throughout the New Testament, and nothing has changed. Beyond these few words, this dichotomy is not a subject that can adequately be addressed in these pages, except to permit me the opportunity here to point out that, in the end, it's not something another person can or should answer for you, but something you can only truly find out for yourself.

113. Jean-Jacques Rousseau (1712–1778) was a famous European philosopher whose work, *The Social Contract*, proposing power in the hands of the people and the "rule of law," had a direct effect on the American and French Revolutions.

114. Realizing that whoever was actually thinking about this kind of stuff would be sitting there forever, the French artist Rodin made a sculpture of the guy. It is on display at the Rodin Museum in Paris, France.

115. Just think of all the people in Rousseau's time who must have been kicking themselves when they realized that instead of just assuming that everyone else already knew they existed, they could have made a fortune by going around announcing that they had a thought from time to time and knew they were alive.

116. This is uniquely illustrated in Nehemiah 8:1-12. In summary we see an instance where "all the people gathered as one man as Ezra and the Levites explained the law. They read from … the law of God, translating to give the sense so that they *understood* the reading. All the people were weeping when they heard the words of the law" (emphasis added). Once they *understood the meaning of the law,* they realized they had not come anywhere close to being able to keep it. But *"Nehemiah … said …* 'this day is holy … do not mourn or weep…. Go, eat of the fat, drink of the sweet [party] and send portions to him who has nothing prepared; for this day

is holy to our Lord. Do not be grieved, for the *joy* of the LORD is your strength.' So … all the people went away to eat, to drink, to send portions and to celebrate a great (party) festival *because they understood* the words which had been made known to them" (emphasis added). Clearly, God was not mad at the people for falling short but, quite the contrary, was very happy because He had been able to put information as to what is good (positive) and what isn't into the minds, hearts, and the energy that flows between human beings.

117. The idea that all human beings share in the unity of this energy flow is a primary concept shared by many of the principal religions throughout history. We noted the Hindu and Buddhist concept of "Om" or "Aum," which is said to describe the sound and "oneness" of this energy flow in the endnotes to chapter 28. Deuteronomy 6:4 proclaims, "Hear, O Israel! The Lord our God is One." See also Mark 12:29.
118. The Greek word for "working" in this passage is *energeo,* from which we get our word "energy."
119. Luke 23:34.
120. See also 1 John 4:7,17-18 MSG.
121. See also John 8:15, 12:47; Romans 8:1.
122. Second Corinthians 11:3 KJV.

WinePressPublishing
Great Books, Defined.

To order additional copies of this book call:
1-877-421-READ (7323)
or please visit our website at
www.WinePressbooks.com

If you enjoyed this quality custom-published book,
drop by our website for more books and information.

www.winepresspublishing.com
"Your partner in custom publishing."

CPSIA information can be obtained at www.ICGtesting.com
Printed in the USA
BVOW081659210313

316134BV00002B/33/P